Selected Titles in This Series

21 **Jürgen Jost,** Bosonic Strings: A Mathematical Treatment, 2001

20 **Lo Yang and S.-T. Yau, Editors,** First International Congress of Chinese Mathematicians, 2001

19 **So-Chin Chen and Mei-Chi Shaw,** Partial Differential Equations in Several Complex Variables, 2001

18 **Fangyang Zheng,** Complex Differential Geometry, 2000

17 **Lei Guo and Stephen S.-T. Yau, Editors,** Lectures on Systems, Control, and Information, 2000

16 **Rudi Weikard and Gilbert Weinstein, Editors,** Differential Equations and Mathematical Physics, 2000

15 **Ling Hsiao and Zhouping Xin, Editors,** Some Current Topics on Nonlinear Conservation Laws, 2000

14 **Jun-ichi Igusa,** An Introduction to the Theory of Local Zeta Functions, 2000

13 **Vasilios Alexiades and George Siopsis, Editors,** Trends in Mathematical Physics, 1999

12 **Sheng Gong,** The Bieberbach Conjecture, 1999

11 **Shinichi Mochizuki,** Foundations of p-adic Teichmüller Theory, 1999

10 **Duong H. Phong, Luc Vinet, and Shing-Tung Yau, Editors,** Mirror Symmetry III, 1999

9 **Shing-Tung Yau, Editor,** Mirror Symmetry I, 1998

8 **Jürgen Jost, Wilfrid Kendall, Umberto Mosco, Michael Röckner, and Karl-Theodor Sturm,** New Directions in Dirichlet Forms, 1998

7 **D. A. Buell and J. T. Teitelbaum, Editors,** Computational Perspectives on Number Theory, 1998

6 **Harold Levine,** Partial Differential Equations, 1997

5 **Qi-keng Lu, Stephen S.-T. Yau, and Anatoly Libgober, Editors,** Singularities and Complex Geometry, 1997

4 **Vyjayanthi Chari and Ivan B. Penkov, Editors,** Modular Interfaces: Modular Lie Algebras, Quantum Groups, and Lie Superalgebras, 1997

3 **Xia-Xi Ding and Tai-Ping Liu, Editors,** Nonlinear Evolutionary Partial Differential Equations, 1997

2.2 **William H. Kazez, Editor,** Geometric Topology, 1997

2.1 **William H. Kazez, Editor,** Geometric Topology, 1997

1 **B. Greene and S.-T. Yau, Editors,** Mirror Symmetry II, 1997

Bosonic Strings: A Mathematical Treatment

AMS/IP

Studies in Advanced Mathematics

Volume 21

Bosonic Strings: A Mathematical Treatment

Jürgen Jost

American Mathematical Society • International Press

Shing-Tung Yau, Managing Editor

2000 *Mathematics Subject Classification*. Primary 81T30; Secondary 83E30, 81T50, 58D30, 32G15, 53A10.

Library of Congress Cataloging-in-Publication Data

Jost, Jürgen, 1956–
Bosonic strings : a mathematical treatment / Jürgen Jost.
p. cm. — (AMS/IP studies in advanced mathematics, ISSN 1089-3288 ; v. 21)
Includes bibliographical references and index.
ISBN 0-8218-2644-1 (alk. paper)
1. String models. 2. Superstring theories. I. Title. II. Series.

QA794.6.S5 J67 2001
530.14—dc21 00-067543

Printed in the United States of America.

∞ The paper used in this book is acid-free and falls within the guidelines established to ensure permanence and durability.

Visit the AMS home page at URL: http://www.ams.org/

Visit the International Press home page at URL: http://www.intlpress.com/

10 9 8 7 6 5 4 3 2 1 06 05 04 03 02 01

Contents

Preface

In classical Newtonian mechanics, in the absence of external forces, a point particle moves along the shortest curve between its initial and final positions, and so its position is determined at all intermediate times. This is no longer so in quantum mechanics, and in Feynman's interpretation, the particle can move along any path between the observed initial and final positions, and so the intermediate positions are not determined anymore. However, we may assign a probability density to each path and this probability density is the higher, the shorter the path is, or, more abstractly, the smaller its action is. On the basis of this probability density, one then attempts to construct a measure on the space of all paths connecting the initial and final positions of the particle, and to integrate over that space with respect to that measure, in order to construct a partition function for evaluating the transition probabilities between the two observed positions. As the space of all connecting paths is infinite dimensional, this construction leads to mathematical problems. It can be made rigorous as a Wiener path integral in the Euclidean situation that we have tacitly assumed in the foregoing discussion. The physical situation of a particle in Minkowski space or a Lorentz manifold is more subtle as it involves oscillatory integrals.

In any case, the Feynman path integral approach is conceptually very appealing, and even in situations where it cannot be made mathematically rigorous, it can be the source of valuable physical as well as mathematical insights.

In order to overcome the difficulties in unifying the electromagnetic, weak, and strong forces on one hand with the gravitational one on the other hand, string theory proposes to treat particles not as points, but to assign them some internal structure, to consider them as strings, i.e. one-dimensional vibrating objects, whose excitation states then correspond to the observed particles and fields. It also includes supersymmetry as a duality between those particles and fields.

If such a string then moves in space-time, it does not traverse a curve, but sweeps out a surface. Again, quantum mechanically, this surface is not determined, but we should construct a probability density on the space of all surfaces connecting the initial and final positions of the string, so that the probability density of any such surface is the higher the smaller the action of the surface is. In analogy to the action for curves which is given by their length, here we should then take the area of a surface.

Thus, the mathematical problem is to construct this probability density and to give a rigorous definition of the resulting functional integral over the space of all surfaces connecting the two string positions. On a conceptual level, this two-dimensional picture has the advantage that in contrast to interactions between point particles that are modeled by intersections of the curves traversed by them and thus represent singularities and lead to the hierarchy of Feynman diagrams, in string theory, interactions between strings are not localized, but simply change the topological type of the traversed surfaces.

In any case, the mathematical problem to construct that functional integral sounds difficult. It can be solved, however. Most treatments are based on studying representations of the diffeomorphism group of the unit circle - which represents the abstract string -, and the Virasoro algebra plays a fundamental rôle. This is not the approach taken here, however. A basic aspect of string theory is that the action is invariant under reparametrizations of the string, in addition to isometries of the space-time in which it moves. The idea then is to systematically divide out these invariances before attempting to define the functional integral. It then turns out that in the end only finitely many degrees of freedom remain. These degrees of freedom are given by the different conformal structures of the Riemann surfaces that can carry the string. (The conformal structures as some of the invariances are broken by considering the Dirichlet integral (Polyakov action) instead of the area (Nambu-Goto action) which preserves the physical features but leads to a more amenable mathematical structure.) Therefore, in the end we can define the partition function for string theories that evaluates the transition probabilities between different positions as an integral over some finite dimensional space, a space of Riemann surfaces of some bounded genus. (As we are not admitting all possible genera simultaneously here, we are truncating the possible interactions between strings, and this point should merit further mathematical study.)

Thus, our approach is based on global Riemannian geometry instead of representation theory. In short, it can be described as the quantization of Plateau's problem for minimal surfaces.

This approach has originally been developed in collaboration with Sergio Albeverio, Sylvie Paycha, and Sergio Scarlatti, and this led to the monograph by the four of us, "A Mathematical Introduction to String Theory", published by Cambridge University Press in 1997. It carried the subtitle "Variational problems, geometric and probabilistic methods", and this blend of methods and combination of diverse approaches that was the result of our stimulating collaboration provided a comprehensive view of the mathematical aspects of string theory. Stimulated by the interest in string theory among scientists at the Max Planck Institute for Mathematics in the Sciences and at the Universities of Halle and Leipzig, I was motivated to rework the geometric approach to string theory, to present additional mathematical connections, and to discuss some more recent developments at the expense, however, of leaving out the probabilistic aspects). The result is the present book.

In the first chapter, we discuss the theory of quantum mechanical point particles in such a manner as to make the analogies with and differences to string theory

transparent, and to introduce some techniques needed in string theory at a simpler example first. The second chapter then develops the theory of Bosonic strings along the lines indicated above. We carefully present the background material from the relevant mathematical fields so as to make the book accessible without presupposing specialized mathematical knowledge. Of course, for some details and proofs, we need to refer to the literature, but it is a basic aim of this book to explain the relevant concepts.
As the title "Bosonic strings" already indicates, we are not incorporating supersymmetry. This is clearly a limitation, and in fact, among the several versions of string theory, the Bosonic string theory is the only one that is not supersymmetric and therefore not acceptable from a physical point of view as its ground state has negative squared mass and is tachyonic. A treatment of superstring theory in this framework will thus have to wait for another occasion. In any case, we do not develop the physical aspects related to the attempted unification of all known fundamental physical interactions. Actually, even from a mathematical point of view, we cover only a small portion of the many aspects and facets of string theory.

Nevertheless, I hope that this book provides a mathematical introduction to string theory that is both accessible and solid, and that it will enable the reader to proceed to the deeper or more recent aspects of the subject, in particular to the so-called M-theory that among other things incorporates all known versions of superstring theory in a unified picture.

It is my pleasure to thank Antje Vandenberg for the competent and efficient TEX-work.

Chapter 1

Point particles

1.1 Point particles and path integrals

We consider a relativistic point particle $x(\tau)$ of mass m, moving in Minkowski space $\mathbb{R}^{d-1,1}$, with metric $\eta_{\mu\nu}$ of signature $(+,...,+,-)$. Its action is given by the path length

$$S = -m \int ds \ , \tag{1.1.1}$$

with

$$ds = \sqrt{-\dot{x}^2} d\tau \ . \tag{1.1.2}$$

Here, $\dot{x}^\mu = \frac{dx^\mu}{d\tau}$, and a summation convention is understood here and in the sequel, i.e.

$$\dot{x}^2 = \eta_{\mu\nu}\dot{x}^\mu \dot{x}^\nu \qquad (x = (x^0, ..., x^{d-1}))$$

τ represents an arbitrary parametrization of the path of the particle, not necessarily by time, and the action does not depend on the choice of this parametrization. In other words, we may perform a reparametrization

$$\tau \mapsto \sigma(\tau)$$

and compute

$$\sqrt{-\left(\frac{dx}{d\tau}\right)^2} \, d\tau = \sqrt{-\left(\frac{dx}{d\sigma}\frac{d\sigma}{d\tau}\right)^2} \cdot \frac{d\tau}{d\sigma} d\sigma = \sqrt{-\left(\frac{dx}{d\sigma}\right)^2} \, d\sigma$$

to see the invariance of the action. Physicists usually only consider infinitesimal symmetries. For that purpose, we look at a smooth family of parametrizations

$$\tau_s(\tau), \ s \in (-\epsilon, \epsilon) \text{ for some } \epsilon > 0,$$

with

$$\tau_0(\tau) = \tau$$

and put

$$\eta := \eta(\tau) := \frac{d\tau_s}{ds}(\tau)_{|_{s=0}}.$$

Then

$$\frac{dx^\mu(\tau_s(\tau))}{ds}_{|s=0} = \eta\, \dot{x}^\mu .$$

With the physicists' notation

$$\delta f := \frac{df(\tau_s(\tau))}{ds}_{|s=0} ,$$

we may thus write the infinitesimal version of the variation of the parametrization as

$$\begin{aligned} \delta\tau &= \eta \\ \delta x^\mu &= \eta \dot{x}^\mu , \end{aligned} \tag{1.1.3}$$

and we have

$$\delta S = 0$$

if η vanishes at the endpoints (i.e. x is parametrized on $[\tau_0, \tau_1], x(\tau_0) = x_0, x(\tau_1) = y_0$, and $\eta(\tau_0) = 0 = \eta(\tau_1)$). In the Feynman path integral version of quantum mechanics, one computes the probability density of a particle with action S (which for the moment need not be of the particular form (1.1.1)) moving from a point x_0 to a point y_0 as

$$P(x_0, y_0) = |\int_{x_0}^{y_0} Dx\, e^{\frac{i}{\hbar}S(x)}\,|^2 \tag{1.1.4}$$

where one attempts to integrate over all paths $x(\tau)$ going from x_0 to y_0, and each such path is weighted by $e^{\frac{i}{\hbar}S(x)}$. It is a nontrivial question to define this infinite dimensional integral rigorously. If one performs a so-called Wick rotation from Minkowski to Euclidean space, amounting to replacing $e^{\frac{i}{\hbar}S(x)}$ by $e^{-\frac{1}{\hbar}S(x)}$, this can be put on a solid mathematical footing, for appropriate actions S, as a Wiener path integral. In such an integral

$$\int_{x_0}^{y_0} Dx\, e^{-\frac{1}{\hbar}S(x)} \tag{1.1.5}$$

the weight of a path becomes exponentially smaller as its action increases, and in the limit $\hbar \to 0$, only the paths minimizing S contribute. Thus, the classical picture is recovered as a limit of the quantum mechanical picture for $\hbar \to 0$. In any case, the measure Dx on the space of all paths from x_0 to y_0 is constructed through finite dimensional approximations. If the path x from x_0 to y_0 goes through an intermediate point z_0, and if the portion from x_0 to z_0 is denoted by x', the one from z_0 to y_0 by x'', then

$$S(x) = S(x') + S(x''),$$

hence

$$\int_{x_0}^{y_0} Dx\, e^{\frac{i}{\hbar}S(x)} = \int dz_0 \int_{x_0}^{z_0} Dx'\, e^{\frac{i}{\hbar}S(x')} \int_{z_0}^{y_0} Dx''\, e^{\frac{i}{\hbar}S(x'')},$$

and hence the idea is to approximate the path integral through repeated subdivisions of the interval from x_0 to y_0 and to replace derivatives in $S(x)$ by difference quotients on all subintervals.
In particular, the measure Dx, if it can be rigorously defined in this manner, is invariant under reparametrizations of the paths x. If the action $S(x)$ happens to be reparametrization invariant as well, then the path integral necessarily diverges because all the infinitely many different parametrizations of the same path yield the same contribution.
In order to overcome these problems and avoid this overcounting, we need some prescription to select some fixed parametrization for each path, and then just discard all the others. This can be viewed as some kind of gauge fixing. Of course, we need to be careful so that the result does not depend on our choice of parametrization.
For our Lagrangian

$$L = -m\sqrt{-\dot{x}^2}, \tag{1.1.6}$$

we have the canonical conjugate variables

$$p_\mu = \frac{\partial L}{\partial \dot{x}^\mu} = \frac{m\dot{x}^\mu}{\sqrt{-\dot{x}^2}}. \tag{1.1.7}$$

We observe a constraint among the canonical momenta

$$p^2 + m^2 = 0. \tag{1.1.8}$$

Furthermore, the Hamiltonian vanishes identically:

$$H := p\dot{x} - L = 0. \tag{1.1.9}$$

Another problem is that the square root in the Lagrangian prevents us from computing functional integrals as Gaussian integrals. Finally, the action S does not extend to the case of massless particles in a meaningful manner.
In order to overcome these problems we introduce an auxiliary variable $\lambda(\tau)$ as a Lagrange multiplier in order to enforce the constraint (1.1.8) and consider

$$L_0 = \frac{1}{\lambda(\tau)}\dot{x}^2(\tau) - \lambda(\tau)m^2. \tag{1.1.10}$$

The corresponding action

$$S_0 = \int d\tau \left(\frac{1}{\lambda(\tau)} \dot{x}^2(\tau) - \lambda(\tau) m^2 \right) \tag{1.1.11}$$

is invariant w.r.t variations

$$\begin{aligned} \delta\tau &= \eta \\ \delta x^\mu &= \eta \dot{x}^\mu \\ \delta\lambda &= \frac{d}{d\tau}(\eta\lambda) \end{aligned} \tag{1.1.12}$$

with η vanishing at the endpoints (i.e. we suppose that x is parametrized on the interval $[\tau_0, \tau_1]$, and we assume $\eta(\tau_0) = 0 = \eta(\tau_1)$). The local variation of λ would be

$$\lambda(\tau_s(\tau)) \frac{d\tau_s}{d\tau}; \tag{1.1.13}$$

namely in that case

$$\frac{d}{ds} \left(\lambda(\tau_s(\tau)) \frac{d\tau_s}{d\tau} \right)_{|s=0} = \frac{d}{d\tau}(\lambda\eta) \qquad \text{in our previous notations.}$$

Thus, λ transforms like a one-dimensional metric density. Similarly, the variations of τ and x^μ in (1.1.12) are simply infinitesimal versions of a reparametrization of the path $x(\tau)$, as

$$\frac{d}{ds} x(\tau_s(\tau))_{|s=0} = \dot{x}(\tau) \frac{d\tau_s}{d\tau}_{|s=0} = \dot{x}\eta$$

if

$$\delta\tau = \frac{d\tau_s}{d\tau}_{|s=0} = \eta.$$

The computation for the invariance of S_0 goes as follows:

$$\begin{aligned} \delta S_0 &= \int d\tau \left(\frac{1}{\lambda} 2\dot{x}\eta\ddot{x} + \frac{1}{\lambda} 2\dot{x}^2\dot{\eta} - \frac{1}{\lambda}\dot{\eta}\dot{x}^2 - \frac{1}{\lambda^2}\eta\dot{\lambda}\dot{x}^2 - \frac{d}{d\tau}(\lambda\eta)m^2 \right) \\ &= \int d\tau \left(\frac{d}{d\tau}\left(\frac{\dot{x}^2}{\lambda}\right)\eta + \dot{\eta}\frac{\dot{x}^2}{\lambda} - \frac{d}{d\tau}(\lambda\eta)m^2 \right) \\ &= \left[\eta\frac{\dot{x}^2}{\lambda} - \lambda\eta m^2 \right]_{\tau_0}^{\tau_1} = 0. \end{aligned}$$

S_0 is a function of the fields $x(\tau)$ and $\lambda(\tau)$, and corresponding Euler-Lagrange equations are obtained as follows:

$$\frac{\delta S_0}{\delta x^\mu} = 0 \quad \text{yields} \quad \frac{d}{d\tau}\left(\frac{\dot{x}^\mu}{\lambda}\right) = 0$$

and

$$\frac{\delta S_0}{\delta \lambda} = 0 \quad \text{yields} \quad \dot{x}^2 + \lambda^2 m^2 = 0.$$

The second equation is an algebraic equation for λ, yielding

$$\lambda = \frac{1}{m}\sqrt{-\dot{x}^2} \tag{1.1.14}$$

and inserting this into the first equation, we obtain

$$m\frac{d}{d\tau}\frac{\dot{x}^\mu}{\sqrt{-\dot{x}^2}} = 0 \tag{1.1.15}$$

which is precisely the Euler-Lagrange equation for S. By reparametrization invariance, we may impose the gauge fixing condition

$$\lambda = \frac{1}{m}, \tag{1.1.16}$$

and inserting this into (1.1.15) yields

$$\ddot{x} = 0. \tag{1.1.17}$$

This was a bit rash, however, because now we have lost the constraint (1.1.8), and thus we have too many solutions. Rather, we must also use the gauge fixed equation of motion for λ as a constraint, i.e.

$$0 = \dot{x}^2 + \lambda^2 m^2 = \dot{x}^2 + 1.$$

In this gauge, the parameter τ becomes the proper time. (In the massless case, we obtain $\dot{x}^2 = 0$, i.e. light like geodesics.) Imposing this gauge in the quantization leads to the so-called Coulomb or light cone quantization. In the Gupta-Bleuler quantization, one imposes the constraint (1.1.8) on state vectors $|\varphi>$. We have in the gauge $\lambda = 1$

$$p_\mu = \frac{\partial S_0}{\partial \dot{x}^\mu} = \eta_{\mu\nu}\frac{dx^\nu}{d\tau},$$

and upon quantization, this becomes

$$p_\mu = -i\frac{\partial}{\partial x^\mu}.$$

Thus the constraint in (1.1.8) leads to the equation

$$\left(-\eta^{\mu\nu}\frac{\partial^2}{\partial x^\mu \partial x^\nu} + m^2\right)\varphi(x) = 0.$$

i.e. the Klein-Gordon equation.

In the preceding, we have considered oscillatory path integrals, namely

$$\int Dx\, e^{\frac{i}{\hbar}S(x)},$$

or

$$\int DxD\lambda\, e^{\frac{i}{\hbar}S_0(x,\lambda)}.$$

As indicated in (1.1.5), it might be easier to work with converging integrals of the type

$$\int DxD\lambda\, e^{-\frac{i}{\hbar}S_1(x,\lambda)}.$$

For that purpose, we switch from Minkowski space $\mathbb{R}^{d-1,1}$ to Euclidean space $\mathbb{R}^d$ by analytic continuation

$$\begin{aligned} x^0(\tau) &\to e^{-i\vartheta}x^0(\tau) \\ \lambda(\tau) &\to e^{-i\vartheta}\lambda(\tau), \end{aligned} \tag{1.1.18}$$

and choosing $\vartheta = \pi$, we replace

$$\begin{aligned} \lambda(\tau) &\to -i\lambda(\tau) \\ x^0(\tau) &\to -ix^d(\tau). \end{aligned} \tag{1.1.19}$$

The density then becomes

$$L_1 = \frac{1}{\lambda(\tau)}\dot{x}^2(\tau) + \lambda(\tau)m^2 \tag{1.1.20}$$

(with $x = (x^1, ..., x^d)$) , and the action

$$S_1 = \int d\tau \left(\frac{1}{\lambda(\tau)}\dot{x}^2(\tau) + \lambda(\tau)m^2\right). \tag{1.1.21}$$

1.2 Faddeev-Popov gauge fixing and BRST symmetry

The Faddeev-Popov procedure is a method to eliminate the integration over orbits of a - usually infinite dimensional - symmetry group from a partition or correlation function. It can be described as follows: Suppose we consider an integral

$$Z := \int e^{-S(A)} dA \tag{1.2.1}$$

where $S(A)$ is some action, and A varies over a configuration space e.g. of random paths or connections. We assume that the action $S(A)$ and the metric on the configuration space underlying the measure dA are invariant under the action of a group $G, A \to gA$ for $g \in G$; the elements of G are called gauge transformations. Because of this invariance under the action of G, we want to eliminate the integration over the orbits of G from the definition of Z, as Z should only count physically inequivalent situations. In order to achieve this, we first choose a slice in the A-space which is transversal to the orbits of G. (This will be possible in our applications below; actually, for the Faddeev-Popov procedure one only needs the existence of local slices). This is called **gauge fixing**. We then write

$$A = (A_1, A_2), \tag{1.2.2}$$

where $A_2 = 0$ corresponds to our slice while for each fixed A_1, A_2 varies on the orbit through $(A_1, 0)$. This change of variables yields a functional determinant via

$$dA = \det \frac{\partial A}{\partial(A_1, A_2)} dA_1 dA_2; \tag{1.2.3}$$

$\det\left(\frac{\partial A}{\partial(A_1,A_2)}\right)$ is called the **Faddeev-Popov determinant**. One finally redefines Z as

$$Z := \int e^{-S(A)} \det \frac{\partial A}{\partial(A_1, A_2)} dA_1, \tag{1.2.4}$$

i.e. one discards the integration over the orbits of G, because each point on an orbit is already represented by a physically equivalent point on the gauge slice. In this manner, Z becomes independent of the choice of local slice and of the parameter A_2 on the orbits of the symmetry group.

In applications, however, as in string theory in Chapter 2, it may happen that while the classical action $S(A)$ is invariant under the action of the group G, the measure dA_1 on the slice is not. This leads to a problem when going from the classical to the quantized theory, and this phenomenon is called anomaly.

We now wish to present a formal procedure expressing the functional determinant in (1.2.3) as an integral over Grassmann variables.

The δ-functional by definition satisfies the following identity:

$$\int dy\ \delta(y)\ f(y) = f(0) \quad \text{for any continuous function } f. \tag{1.2.5}$$

If $y = h(x)$, for an invertible function h, we get

$$\int dx\ \delta(h(x))\ |\frac{\partial h}{\partial x}| f(h(x)) = f(0), \tag{1.2.6}$$

or, with $\varphi(x) := f(h(x))$

$$\int dx\ \delta(h(x))|\frac{\partial h}{\partial x}|\varphi(x) = \varphi_{|h(x)=0}. \tag{1.2.7}$$

Here, x can be a vector variable $x = (x^1, ..., x^d)$, and (1.2.7) can then be written as

$$\int \prod_{i=1}^{d} dx^i\ \delta(h^i(x))|\det \frac{\partial h^j}{\partial x^k}|\varphi(x) = \varphi_{|h(x)=0}. \tag{1.2.8}$$

For the sequel, we shall now express the delta functional $\delta(h^i(x))$ and the determinant $\det \frac{\partial h^j}{\partial x^k}$ through integrals: First

$$\prod_{i=1}^{d} \delta(h^i(x)) = \prod_{i} \int_0^{2\pi} \frac{d\mu_i}{2\pi}\ e^{-i\mu_i\, h^i(x)}. \tag{1.2.9}$$

For the representation of the determinant, we need to introduce Grassmann variables; or slightly more generally for the sequel, Clifford variables. We thus assume that we have a Clifford algebra generated by $\vartheta_1, ..., \vartheta_n$ and $\bar{\vartheta}_1, ..., \bar{\vartheta}_n$ satisfying

$$\begin{aligned} [\vartheta_i, \vartheta_j]_+ &:= \vartheta_i\vartheta_j + \vartheta_j\vartheta_i = 0 \\ [\bar{\vartheta}_i, \bar{\vartheta}_j]_+ &= 0 \\ [\vartheta_i, \bar{\vartheta}_j]_+ &= \mu\delta_{ij} \quad \text{for some } \mu \in \mathbb{R} \end{aligned} \tag{1.2.10}$$

(in the Grassmann case, $\mu = 0$, but in the sequel, we shall need $\mu = 1$).
For a function $f(\vartheta_1, \bar{\vartheta}_1, ..., \vartheta_n, \bar{\vartheta}_n)$, the integral

$$\int d\vartheta_1 d\bar{\vartheta}_1\ ...\ d\vartheta_n d\bar{\vartheta}_n\ f(\vartheta_1, ..., \bar{\vartheta}_n)$$

then is defined as the coefficient of $\bar{\vartheta}_n\vartheta_n\ ...\ \bar{\vartheta}_1\vartheta_1$ in the expansion of f. Gaussian integrals with Grassmann variables can then be easily evaluated

$$I(A) := \int d\vartheta_1 d\bar{\vartheta}_1\ ...\ d\vartheta_n d\bar{\vartheta}_n\ \exp\left(\sum_{i,j=1}^{n} \bar{\vartheta}_i A_{ij} \vartheta_j\right)$$

where $A = (A_{ij})_{i,j=1,\dots,n}$ is a matrix with real or complex entries. The expansion of the exponential function is

$$\exp\left(\sum_{i,j=1}^{n} \bar{\vartheta}_i A_{ij} \vartheta_j\right) = \prod_{i,j=1}^{n} \exp\left(\bar{\vartheta}_i A_{ij} \vartheta_j\right)$$

$$(1.2.11) \qquad = \prod_{i,j=1}^{n} \left(1 + \bar{\vartheta}_i A_{ij} \vartheta_j\right)$$

from the usual power series for the exponential function and the nilpotency of the $\vartheta_i, \bar{\vartheta}_i$.

Expanding the product, we see that the terms contributing to the coefficient of $\bar{\vartheta}_n \vartheta_n \dots \bar{\vartheta}_1 \vartheta_1$ in the expansion are precisely the following

$$\sum_{\substack{\text{permutations} \\ (j_1,\dots,j_n)\text{of}(1,\dots,n)}} A_{n,j_n} A_{n-1,j_{n-1}} \dots A_{1,j_1} \bar{\vartheta}_n \vartheta_{j_n} \dots \bar{\vartheta}_1 \vartheta_{j_1}.$$

$$= \det A \; \bar{\vartheta}_n \vartheta_n \bar{\vartheta}_{n-1} \vartheta_{n-1} \dots \bar{\vartheta}_1 \vartheta_1$$

by permuting the ϑ_j and using the anticommutation relations. Consequently, we obtain

$$\int d\vartheta_1 \dots d\bar{\vartheta}_n \exp\left(\sum \bar{\vartheta}_i A_{ij} \vartheta_j\right) = \int d\vartheta_1 d\bar{\vartheta}_1 \dots d\vartheta_n d\bar{\vartheta}_n \prod (1 + \bar{\vartheta}_i A_{ij} \vartheta_j)$$

$$(1.2.12) \qquad = \int d\vartheta_1 d\bar{\vartheta}_1 \dots d\vartheta_n d\bar{\vartheta}_n \det A \bar{\vartheta}_n \vartheta_n \dots \bar{\vartheta}_1 \vartheta_1$$

$$= \det A.$$

Combining (1.2.8), (1.2.9) and (1.2.12), we obtain the formula

(1.2.13)

$$\varphi|_{h(x)=0} = \int \prod_{i=1}^{d} dx^i \frac{d\mu_i}{2\pi} d\vartheta_1 d\bar{\vartheta}_1 \dots d\vartheta_1 d\bar{\vartheta}_1 \; \varphi(x) \exp\left(-i\mu_i h^i(x) + \bar{\vartheta}_\gamma \frac{\partial h^\beta}{\partial x^\gamma} \vartheta_\beta\right).$$

In our situation, we are considering a functional integral

$$\int dx \exp(-S_1(x)),$$

and we assume that we are fixing the gauge through the equation $H(x) = 0$. Implementing the above procedure, we thus replace this functional integral by the one obtained through Faddeev-Popov gauge fixing as just described, i. e.

$$(1.2.14) \qquad \int dx \; d\mu \; d\vartheta \; d\bar{\vartheta} \; \exp\left(-S_1(x) - i\mu h(x) + \bar{\vartheta} \frac{\partial h(x)}{\partial x} \vartheta\right)$$

(omitting all indices, as well as the factor $\frac{1}{2\pi}$ which can be accounted for by a redefinition of the integral). We should note that the meaning of x in (1.2.14) is different from the one in (1.2.13). Namely, in (1.2.13), x represents only the gauge degrees of freedom, whereas in (1.2.14) it represents all independent variables. If the Lie algebra of the gauge group is generated by $X_\gamma, \gamma = 1, ..., n$, and if we write $X_\gamma(x) =: \delta_\gamma x$, then $\frac{\partial h^\beta}{\partial x^\gamma}$ has to be replaced by $\frac{\partial h^\beta}{\partial x^i}\delta_\gamma x^i$.

Thus, in addition to the original gauge-invariant action S_1, we have a gauge fixing action $i\mu h(x)$, with Lagrange multipliers μ^α, and a Faddeev-Popov action $-\bar{\vartheta}\frac{\partial h}{\partial x}\vartheta$, with anticommuting "ghosts" $\vartheta_\beta, \bar{\vartheta}_\gamma$. The idea of the Faddeev-Popov method then is to treat $\mu, \vartheta, \bar{\vartheta}$ as new variables on the same footing with x. Now the motivation for the Faddeev-Popov procedure was to eliminate the gauge redundancy from the path integral, and we should clarify how this is achieved. We thus assume that S_1 is invariant under some closed Lie algebra $\mathfrak{g}$, with generators $X_1, ..., X_N$, satisfying

$$[X_i, X_j] = f_{ij}^k X_k, \tag{1.2.15}$$

where the f_{ij}^k are the structure constants of $\mathfrak{g}$. We shall also write w^i in place of $\bar{\vartheta}, b_j$ in place of ϑ_j, and assume that the "antighosts" b_j transform in the adjoint representation of $\mathfrak{g}$, while the "ghosts" w^i transform in the dual of the adjoint representation. If $\mathfrak{g}$ is considered as the tangent space T_eG of a Lie group at the identity, then w^i can be considered as a 1-form, b_j as a tangent vector to G at e. The exterior derivative for $\mathfrak{g}$ or G for the cohomology with values in the representation R defined by the X_j is given by

$$\begin{aligned} d(\varphi w^{i_1} \wedge ... \wedge w^{i_n}) =&(X_j\varphi)w^j \wedge w^{i_1} \wedge ... \wedge w^{i_n} \\ &- \frac{1}{2}(-1)^j f_{kl}^{i_j} w^k \wedge w^l \wedge w^{i_1} \wedge ... \wedge \widehat{w^{i_j}} \wedge ... \wedge w^{i_n}, \end{aligned} \tag{1.2.16}$$

for φ a function on the representation space defined by the X_j (where $\widehat{w^m}$ means that w^m is omitted).

$$d = w^j X_j - \frac{1}{2} f_{kl}^j w^k\, w^l\, b_j. \tag{1.2.17}$$

In the physics literature, this operator is called the **BRST** (Becchi-Rouet-Stora-Tyupin) **operator** and is denoted by Q, and in the sequel, we shall also utilize that notation. The important property is that d or Q is nilpotent, i.e. satisfies

$$Q^2 = 0. \tag{1.2.18}$$

We have the anticommutation relation

$$[w^i, b_j]_+ = \delta_j^i \tag{1.2.19}$$

so to each w^1, we assign the ghost number $+1$, and to each b_j the ghost number -1. Thus, Q raises the ghost number by 1, and the ghost number in mathematical terminology counts the degree of a diffential form. The total ghost number is defined as

$$U := \sum_j w^j\, b_j. \tag{1.2.20}$$

The corresponding cohomology groups $H^k(\mathfrak{g}, R)$ are called the **BRST cohomology groups**. States $|\varphi>$ are sections of $C^\infty(R) \otimes \wedge^n(\mathfrak{g}^*) \cong Hom(\wedge^n(\mathfrak{g}), C^\infty(R))$. Physical states are gauge independent and thus satisfy

$$Q|\varphi>= 0$$

States $|\varphi>$ and $|\varphi'>$ with

$$|\varphi> - |\varphi'>= Q|\psi>,$$

i.e. states differing only by a cohomologically trivial state are physically equivalent. In other words, physical states are BRST cohomology classes. Q acts on the ghosts and antighosts by anticommutators:

$$\begin{aligned} \delta w^i &= [Q, w^i]_+ = -\frac{1}{2} f^i_{kl} w^k\, w^l \\ \delta b_i &= [Q, b_i]_+ = X_i - f^k_{ij} w^j\, b_k. \end{aligned} \tag{1.2.21}$$

The point of the BRST formalism, however, is that it mixes the ghost degrees of freedom with the gauge degrees of freedom so that they can compensate each other. For that purpose, we introduce an anticommuting parameter ϵ and define the BRST transformation as

$$\begin{aligned} \delta x^i &= i\epsilon\bar{\vartheta}^\gamma \delta_\gamma x^i, \qquad \text{with} \qquad \delta_\gamma x^i := X_\gamma(x^i) \\ \delta\mu_\alpha &= 0 \\ \delta\vartheta_\alpha &= \varepsilon\mu_\alpha \\ \delta\bar{\vartheta}^\alpha &= \frac{i}{2}\varepsilon f^\alpha_{\beta\gamma}\bar{\vartheta}^\beta\bar{\vartheta}^\gamma. \end{aligned} \tag{1.2.22}$$

This implies that

$$\begin{aligned} \delta(\vartheta_\alpha h^\alpha) &= i\varepsilon\left(-i\mu_\alpha h^\alpha \;-\; \vartheta_\alpha \frac{\partial h^\alpha}{\partial x^i}\delta_\gamma x^i \bar{\vartheta}^\gamma\right) \\ &= i\varepsilon\left(-i\mu_\alpha h^\alpha \;+\; \bar{\vartheta}^\gamma \frac{\partial h^\alpha}{\partial x^i}\delta_\gamma x^i \vartheta_\alpha\right) \end{aligned} \tag{1.2.23}$$

if the $\bar{\vartheta}^\gamma$ anticommute (Here $\frac{\partial h^\alpha}{\partial x^i}\delta_\gamma x^i$ stands for what formerly was $\frac{\partial h^\alpha}{\partial x^\gamma}$, as explained above.) This is precisely the sum of the gauge fixing and the Faddeev-Popov action. Since the BRST operator is nilpotent, i.e. $\delta^2 = 0$, we conclude

$$\delta\left(-i\mu_\alpha h^\alpha + \bar{\vartheta}^\gamma \frac{\partial h^\alpha}{\partial x^i}\delta_\gamma x^i \vartheta_\alpha\right) = 0. \tag{1.2.24}$$

Since S_1 by itself is gauge invariant, the whole Lagrangian density appearing in (1.2.14) is invariant:

$$\delta\left(-S_1 - i\mu h + \bar{\vartheta}\frac{\partial h}{\partial x}\vartheta\right) = 0. \tag{1.2.25}$$

As an exercise, it is also instructive to verify (1.2.24) directly:

$$\begin{aligned}
&\delta\left(-i\mu_\alpha h^\alpha + \bar{\vartheta}^\gamma \frac{\partial h^\alpha}{\partial x^i}\delta_\gamma x^i\,\vartheta_\alpha\right) \\
&= -\varepsilon\mu_\alpha \frac{\partial h^\alpha}{\partial x^i}\delta_\gamma x^i \bar{\vartheta}^\gamma + \frac{i}{2}\varepsilon f^\gamma_{\beta\delta}\bar{\vartheta}^\beta\bar{\vartheta}^\delta \frac{\partial h^\alpha}{\partial x^i}\delta_\gamma x^i \vartheta_\alpha \\
&\quad + \varepsilon\bar{\vartheta}^\gamma \frac{\partial h^\alpha}{\partial x^i}\delta_\gamma x^i \mu_\alpha - i\varepsilon\bar{\vartheta}^\gamma \frac{\partial h^\alpha}{\partial x^i}\bar{\vartheta}^\beta \delta_\beta\delta_\gamma x^i \vartheta_\alpha \\
&\quad - i\varepsilon\bar{\vartheta}^\gamma \frac{\partial^2 h^\alpha}{\partial x^i \partial x^j}\delta_\gamma x^i \delta_\beta x^j \bar{\vartheta}^\beta \vartheta_\alpha.
\end{aligned}$$

Here, the first and the third term cancel each other, while the last one vanishes because $\delta_\gamma x^i \delta_\beta x^j \frac{\partial^2 h^\alpha}{\partial x^i \partial x^j}$ is symmetric in β and γ while $\bar{\vartheta}^\gamma\bar{\vartheta}^\beta$ is antisymmetric. Finally, since $\bar{\vartheta}^\gamma\bar{\vartheta}^\beta$ is antisymmetric, we may write the fourth term as

$$-\frac{i}{2}\varepsilon\bar{\vartheta}^\gamma\bar{\vartheta}^\beta \frac{\partial h^\alpha}{\partial x^i}(\delta_\beta\delta_\gamma x^i - \delta_\gamma\delta_\beta x^i)\vartheta_\alpha,$$

and since $\delta_\beta\delta_\gamma - \delta_\gamma\delta_\beta = f^\eta_{\beta\gamma}\delta_\eta$, this cancels the second term.

1.3 BRST quantization of the point particle

In the preceding §, it was implicitly assumed that we had only finitely many degrees of freedom for the variable $x = (x^1, ..., x^d)$. In the situation considered in §1.1, however, we had a path $x(\tau)$. This path has infinitely many degrees of freedom, namely its values at all the points τ in its interval of definition, and so the discrete index i of x^i has to be replaced by the real variable τ of $x(\tau)$. Obviously, this should be a point of grave concern, but for the present considerations we ignore this and proceed purely formally, the justification being that the path integral

$$\int dx\, d\lambda\, e^{-S_1(x,\lambda)} \qquad \text{(with } S_1 \text{ as in (1.1.21))} \tag{1.3.1}$$

over all paths satisfying the specified boundary conditions is ill-defined anyway, and our goal is to arrive at a meaningful expression for an integral that only counts inequivalent paths through heuristic considerations from (1.3.1). We recall the action

$$S_1(x,\lambda) = \int d\tau \left(\frac{1}{\lambda(\tau)}\dot{x}^2(\tau) + \lambda(\tau)m^2\right) \tag{1.3.2}$$

from (1.1.21). We assume again that the path $x(\tau)$ is parametrized on the interval $I := [\tau_0, \tau_1]$ and has fixed boundary points $x(\tau_0) = p_0, x(\tau_1) = p_1$. Given $\lambda(\tau)$, we write

$$x(\tau) = x_0(\tau) + \xi(\tau),$$

where x_0 satisfies the boundary conditions $x_0(\tau_0) = p_0, x_0(\tau_1) = p_1$ and

$$\frac{1}{\lambda(\tau)} \frac{d}{d\tau} \left(\frac{1}{\lambda(\tau)} \frac{dx_0}{d\tau} \right) = 0,$$

or in abbreviated form, with $^\bullet := \frac{d}{d\tau}$,

$$\frac{1}{\lambda} \left(\frac{\dot{x}_0}{\lambda} \right)^\bullet = 0. \tag{1.3.3}$$

$\xi(\tau)$ then satisfies homogeneous boundary conditions $\xi(\tau_0) = \xi(\tau)$. Moreover

$$S_1(x_0 + \xi, \lambda) = \int d\tau \left(\frac{1}{\lambda} \dot{x}^2 + \frac{1}{\lambda} \dot{\xi}^2 + \lambda m^2 \right), \tag{1.3.4}$$

since

$$\int d\tau \, \frac{1}{\lambda} \dot{x}_0 \dot{\xi} = 0$$

from (1.3.3) and the zero boundary conditions for ξ.
We now work with the Sobolev space $H_0^s(I, \mathbb{R}^d)$ of functions from I to $\mathbb{R}^d$ with zero boundary values and square integrable derivatives up to order s. (A short treatment of Sobolev spaces can be found in § 2.2 below.) Here, all integrals are defined w.r.t. the "volume" form $\lambda(\tau)d\tau$ on I. The topology of the spaces $H_0^s(I, \mathbb{R}^d)$ is independent of the choice of λ, provided λ is sufficiently regular. We also consider the affine spaces

$$F^s := x_0 + H_0^s(I, \mathbb{R}^d)$$

for $1 \le s < \infty$, which are affine Hilbert spaces for $s \in \mathbb{N}$.

$$F^\infty := \bigcap_{s \in \mathbb{N}} F^s$$

is an affine Fréchet space which also carries a so-called strong ILH-structure. For $V, W \in T_x F^1$, we consider the H^1 product

$$(V, W)_\lambda = \int_I V^\alpha(\tau) W^\alpha(\tau) \lambda(\tau) d\tau. \tag{1.3.5}$$

Here, tangent vectors to F^1 at x are identified with vector fields in $\mathbb{R}^d$ along x. For $s = 1$, this induces the Riemann-Hilbert structure on F^1, while for $s \ge 2$ it only introduces a weak Riemannian structure on F^s, and F^s is not complete

anymore w.r.t this topology.
Next, we have the diffeomorphism group

$$\mathcal{D}^\ell := \{\text{diffeomorphisms } \varphi : I \to I \text{ of Sobolov class } H^\ell\} \, (\ell \geq 1).$$

For $\ell = 1$, $\varphi \in H^\ell$ need not be differentiable, and so in that case, we need to consider the completion of the space of diffeomorphisms of class C^∞ w.r.t the H^ℓ norm. These diffeomorphism groups act on the mapping spaces by composition. More precisely, the action is given by

$$\begin{aligned} F^s \times \mathcal{D}^\ell &\to F^t \quad , \quad t \leq \min(\ell, s) \\ (x, \varphi) &\mapsto x \circ \varphi. \end{aligned}$$

This action is of class C^∞ in x, for fixed φ, but only of class C^{s-t} in φ. Namely, if one considers a smooth family $(\varphi_t)_{-\varepsilon<t<\varepsilon}$ in $\mathcal{D}^\ell$ and $x \circ \varphi_t$, then each differentation w.r.t t causes a derivative of x by the chain rule. For $s = t = 1$, the action of $\mathcal{D}^\ell$ is only continuous. The group $\mathcal{D}^\ell$ also acts on the space of metrics on I, via

$$(\lambda, \varphi) \mapsto \varphi^* \lambda$$

with

$$\varphi^* \lambda(\tau) = \lambda(\varphi(\tau)) \, \frac{d\varphi(\tau)}{d\tau}. \tag{1.3.6}$$

If λ is of Sobolev class H^s and $\varphi \in \mathcal{D}^\ell$, then $\varphi^* \lambda$ is of class H^t for $t \leq \min(s, \ell - 1)$ which is easily computed from the Sobolev embedding theorem. Again, there is a loss of regularity caused by the chain rule as above.

Lemma 1.3.1.

$$(V, W)_\lambda = (V \circ \varphi, W \circ \varphi)_{\varphi^* \lambda} \quad \textit{for} \quad \varphi \in \mathcal{D}^\ell. \tag{1.3.7}$$

Thus, the weak Riemannian structure on F^s introduced above is invariant under the action of the diffeomorphism group.

Proof.

$$\begin{aligned} &\int_I V(\varphi(\tau)) \, W(\varphi(\tau)) \, \varphi^* \lambda(\tau) \, d\tau \\ &= \int_I V(\sigma) \, W(\sigma) \, \lambda(\sigma) \, d\sigma \quad \text{with} \quad \sigma = \varphi(\tau), \end{aligned}$$

by the chain rule, which is the required formula. □

In the finite-dimensional case, a Riemannian structure induces a volume form. This is no longer so in infinite dimensions, but we shall nevertheless utilise the

above weak Riemannian structure on F^s as some heuristic guiding principle for our attempts to define integration w.r.t mappings x and metric densities λ. Lemma 1.3.1 shows that the integration w.r.t x is invariant under the action of $\mathcal{D}^\ell$.
We start with the integral

(1.3.8)
$$\begin{aligned} &\int dx\, e^{-S_1(x,\lambda)} \\ =&\int d\xi\, e^{-\int d\tau\left(\frac{1}{\lambda}\dot{x}_0^2+\lambda m^2\right)}e^{-\int d\tau \frac{1}{\lambda}\dot{\xi}^2} \end{aligned}$$

(x_0 in the decomposition $x = x_0 + \xi$ is fixed by the differential equation (1.3.3) and the boundary condition and so does not represent a degree of freedom for the functional integral. Thus, we only integrate w.r.t. ξ). The factor $\exp(-\int d\tau(\frac{1}{\lambda}\dot{x}_0^2 + \lambda m^2))$ does not contain the variable ξ anymore, and so it can be put in front of the integral. This factor contains all the concrete physical information, namely the boundary condition and the mass, whereas the remaining functional integral is universal. This decoupling of the physical information from the functional integral represents a remarkable simplification.
Since by our heuristic arguments, functional interpretation w.r.t. ξ should be gauge independent, we may fix the gauge $\lambda = 1$. This is a crucial step that simplifies our functional integral to become

(1.3.9)
$$\begin{aligned} &e^{-\int d\tau(\dot{x}_0^2+m^2)}\int d\xi\, e^{-\int d\tau\, \dot{\xi}^2} \\ =&e^{-\int d\tau(\dot{x}_0^2+m^2)}\int d\xi\, e^{\int d\tau\, \xi\ddot{\xi}}. \end{aligned}$$

We let $(\xi_\nu)_{\nu\in\mathbb{N}}$ be the sequence of L^2-orthonormal eigenvectors of $\frac{d^2}{d\tau^2}$ on the interval I:

$$\begin{aligned} \frac{d^2}{d\tau^2}\,\xi_\nu \;+\; \sigma_\nu\,\xi_\nu \;&=\; 0, \\ \int_I \xi_\nu(\tau)\xi_\mu(\tau)d\tau \;&=\delta_{\mu\nu} \end{aligned}$$

($\sigma_\nu = \nu\frac{\pi}{\ell}$, $\nu\in\mathbb{N}$, for $\ell :=$ length $(I) = \tau_1 - \tau_0$).

We write $\xi \in H_0^1(I,\mathbb{R}^d)$ as a series

$$\xi = \sum_\nu a_\nu\xi_\nu, \quad \text{with} \quad a_\nu(\xi) = \int_I \xi_\nu(\tau)\xi(\tau)d\tau.$$

Then

(1.3.10)
$$e^{-\int d\tau\dot{\xi}^2} = e^{\int \xi\ddot{\xi}d\tau} = e^{-\sum\limits_\nu \sigma_\nu a_\nu(\xi)^2}.$$

The heuristic idea then is to put

$$d\xi = \prod_{\nu} da_\nu(\xi),$$

with $da_\nu(X)$ being the measure induced from $(.,.)_1$ (see (1.3.5)) on the 1-dimensional subspace of F^1 spanned by ξ_ν. We then recall the formula

$$\int_{\mathbb{R}} e^{-\sigma y^2} dy = \left(\frac{2\pi}{\sigma}\right)^{\frac{1}{2}} \tag{1.3.11}$$

and attempt the definition

$$\int d\xi \; e^{-\int d\tau \dot{\xi}^2} = \left(\det \frac{d^2}{d\tau^2}\right)^{-\frac{d}{2}} \tag{1.3.12}$$

where $\det \frac{d^2}{d\tau^2}$ stand for the formal product of the eigenvalues of the operator $\frac{d^2}{d\tau^2}$ and needs to be defined in a rigorous manner. This will be performed below in § 2.6. The d in the exponent arises from the fact that ξ takes its values in $\mathbb{R}^d$ and thus has d scalar components.

There is some ambiguity in the choice of factors as one sees by comparing (1.3.11) with (1.3.12), as the factor 2π does not appear in (1.3.12) anymore. Of course, the infinite product of 2π with itself cannot be defined in a meaningful manner. However, for certain reasons, it might be desirable to include an additional factor (length $I)^{\frac{d}{2}}$ in the rhs of (1.3.12).

One more remark: (1.3.12) is the integral of a bosonic Gaussian, and so the determinant comes out with a negative exponent, in contrast to the fermionic Gaussian integral (1.2.12) where we get a positive exponent.

We now come to the action of the diffeomorphism group $\mathcal{D}^\ell$ on the space $\mathcal{M}^k$ of metrics $\lambda^2(\tau)d\tau$ on I of Sobolev class k. On $\mathcal{M}^k$, we again have a weak Riemannian metric given by

$$(\gamma_1, \gamma_2)_\lambda := \int_I d\tau \frac{1}{\lambda^3(\tau)} \gamma_1(\tau)\gamma_2(\tau). \tag{1.3.13}$$

Since γ_1, γ_2 transform under the action of $\varphi \in \mathcal{D}^\ell$ in the same manner as $\lambda^2(\tau)d\tau$, i.e.

$$\gamma \mapsto \varphi^*\gamma, \quad \text{with} \quad \varphi^*\gamma(\tau) = \gamma(\varphi(\tau))\dot{\varphi}(\tau)^2,$$

we get the analogue of Lemma 1.3.1:

Lemma 1.3.2.

$$(\gamma_1, \gamma_2)_\lambda = (\varphi^*\gamma_1, \varphi^*\gamma_2)_{\varphi^*\lambda}, \tag{1.3.14}$$

meaning that the weak Riemannian structure on $\mathcal{M}^k$ defined by (1.3.13) is invariant under the action of $\mathcal{D}^\ell$.

We now consider a smooth family

$$(\varphi_t)_{-\varepsilon<t<\varepsilon} \subset \mathcal{D}^\ell, \quad \text{with} \quad \varphi_0 = \text{identity},$$

and put

$$V(\tau) := \frac{d}{dt}\varphi_t(\tau)_{|t=0}. \tag{1.3.15}$$

We then have

$$\begin{aligned} \frac{d}{dt}(\varphi_t^* \lambda^2)_{|t=0} &= \frac{d}{dt}\left(\lambda^2(\varphi_t(\tau))\left(\frac{d\varphi_t(\tau)}{d\tau}\right)^2\right)_{|t=0} \\ &= 2\lambda\dot{\lambda}V + 2\lambda^2\dot{V} \ = 2\lambda^2(\nabla^\lambda V) \text{ where } \nabla^\lambda := \frac{d}{d\tau} + \frac{\dot{\lambda}}{\lambda} \\ &\qquad\qquad\qquad \text{is the covariant derivative w.r.t the metric } \lambda \\ &=: L_V\lambda^2 \end{aligned} \tag{1.3.16}$$

(this is the Lie derivative of the metric $\lambda^2(\tau)d\tau^2$ in the direction of the vector field $V(\tau)$).

We now wish to consider functional integration w.r.t. λ, based on the formal measure induced by (1.3.13). As before, this formal measure is gauge invariant, by Lemma 1.3.2, and we may thus again impose our gauge condition $\lambda = 1$. We decompose $\gamma \in T_{\lambda^2}\mathcal{M}^k$ as

$$\gamma(\tau) = c\lambda^2(\tau) + (L_V\lambda^2)(\tau) \tag{1.3.17}$$

for some constant $c \in \mathbb{R}$ and some vector field $V(\tau)$ on I that vanishes on ∂I. This decomposition is orthogonal w.r.t (1.3.13), namely using (1.3.16)

$$\int_I d\tau \frac{1}{\lambda^3(\tau)}\, c\lambda^2(\tau)2\lambda\frac{d}{d\tau}(\lambda V) = 0$$

if V vanishes at the boundary of I . Conversely, every γ can be represented in the form (1.3.17); namely, we only have to choose

$$c = \frac{1}{2\int\lambda(\tau)d\tau}\int\frac{\gamma(\tau)}{\lambda(\tau)}\,d\tau,$$

as $L_V\lambda^2 = 2\lambda\frac{d}{d\tau}(\lambda V)$ so that every γ_0 with $\int\frac{\gamma_0(\tau)}{\lambda(\tau)}d\tau = 0$ can be represented as $L_V\lambda^2$ for some V. The infinitesimal decomposition (1.3.17) corresponds to the global fact that any parametrized curve with a metric (element of length) along it can be parametrized proportionally to arclength, by a change of parametrization through a diffeomorphism; the constant c in (1.3.17) then is the factor of proportionality.

The vector field V_τ represents an infinitesimal diffeomorphism , see (1.3.15), and our aim is to divide out the action of the diffeomorphism group in our functional integration, and we thus now implement the Faddeev-Popov procedure. We write the variation γ of λ^2 as

$$\gamma = (c, V)$$

with $c \in \mathbb{R}$ and V a vector field on I with vanishing boundary values, according to (1.3.17). Now the Faddeev-Popov prescription is to omit the integration w.r.t the gauge degree of freedom and to introduce an appropriate functional determinant instead. Here, the gauge degree of freedom is represented by V, and from (1.3.17) we see that the functional determinant to be included is

$$2\lambda \det \bigtriangledown^\lambda,$$

$$\text{with} \quad \bigtriangledown^\lambda = \frac{d}{d\tau} + \frac{\dot\lambda}{\lambda}.$$

In the formulation of § 1.2, we thus get the action

$$S(x, \lambda, \mu, \vartheta, \bar\vartheta) \;=\; \int d\tau \left(\frac{1}{\lambda(\tau)}\, \dot x^2(\tau) + \lambda(\tau) m^2 + i\mu(\lambda^2 - 1) + 2\lambda^2 \vartheta \bigtriangledown^\lambda \bar\vartheta \right)$$

corresponding to $h = 1 - \lambda^2$, our standard gauge.
This is invariant under the BRST transformation (cf (1.1.12))

$$\begin{aligned}
\delta\mu &= i\varepsilon\bar\vartheta\dot x \\
\delta\lambda &= i\varepsilon \frac{d}{d\tau}(\bar\vartheta\lambda) \\
\delta\mu &= 0 \\
\delta\vartheta &= \varepsilon\mu \\
\delta\bar\vartheta &= i\varepsilon\bar\vartheta\,\dot{\bar\vartheta}
\end{aligned}$$

(Because of the nilpotency of $\bar\vartheta$, $\bar\vartheta\,\dot{\bar\vartheta} = \bar\vartheta(\dot{\bar\vartheta} + \frac{\dot\lambda}{\lambda}\bar\vartheta)$.) Namely, S_1 is invariant under the transformation (1.1.12), and the remaining contributions are easily checked to cancel; most easily we verify that

$$\begin{aligned}
\delta((1-\lambda^2)\vartheta) &= \varepsilon\mu(1-\lambda^2) + i\varepsilon 2\lambda^2\vartheta \left(\dot{\bar\vartheta} + \frac{\dot\lambda}{\lambda}\bar\vartheta \right) \\
&= i\varepsilon(i\mu(\lambda^2 - 1) + \vartheta \bigtriangledown^\lambda \bar\vartheta)
\end{aligned}$$

and use the nilpotency $\delta^2 = 0$ to ensure that

$$\delta(i\mu(\lambda^2 - 1) + \vartheta \bigtriangledown^\lambda \bar\vartheta) = 0$$

(That the BRST variation is nilpotent is easily checked directly in the present case; if we perform variations with parameters ε and ε', we have for example

$$\begin{aligned}
\delta_{\varepsilon'}\delta_\varepsilon\, \lambda &= \delta_{\varepsilon'}(i\varepsilon(\dot\lambda\bar\vartheta + \lambda\dot{\bar\vartheta})) \\
= -\varepsilon\varepsilon' &\left\{ (\lambda\ddot{\bar\vartheta} + 2\dot\lambda\dot{\bar\vartheta} + \ddot\lambda\bar\vartheta)\bar\vartheta + (\lambda\dot{\bar\vartheta} + \dot\lambda\bar\vartheta)\dot{\bar\vartheta} + \dot\lambda\bar\vartheta\dot{\bar\vartheta} + \lambda\bar\vartheta\ddot{\bar\vartheta} \right\} = 0.
\end{aligned}$$

The preceding simplifies if we fix the gauge $\lambda(\tau) \equiv 0$. The Euler-Lagrange equation of S w.r.t λ is

$$0 = -\frac{1}{\lambda^2}\dot{x}^2 + m^2 + 2i\lambda\mu + 2\lambda(\vartheta\dot{\bar{\vartheta}} - \dot{\vartheta}\bar{\vartheta})$$

and we can use this to eliminate μ:

$$\mu = -i\left(\frac{1}{2}\dot{x}^2 - \frac{1}{2}m^2 - \vartheta\dot{\bar{\vartheta}} + \dot{\vartheta}\bar{\vartheta}\right) \qquad \text{for} \qquad \lambda = 1.$$

Thus, the action becomes

$$\int d\tau(\dot{x}^2(\tau) + m^2 + 2\vartheta\dot{\bar{\vartheta}}),$$

and the BRST variation

$$\begin{aligned}
\delta x &= i\varepsilon\bar{\vartheta}\dot{x}\\
\delta\vartheta &= i\varepsilon(-\frac{1}{2}\dot{x}^2 + \frac{1}{2}m^2 - \dot{\vartheta}\bar{\vartheta} + \vartheta\dot{\bar{\vartheta}})\\
\delta\bar{\vartheta} &= i\varepsilon\bar{\vartheta}\dot{\bar{\vartheta}}
\end{aligned}$$

(In order to check this, we compute the variation of the integrand $\dot{x}^2(\tau) + m^2 + 2\vartheta\dot{\bar{\vartheta}}$ as a total derivative:

$$\begin{aligned}
2i\varepsilon\dot{\bar{\vartheta}}\dot{x}^2 + 2i\varepsilon\bar{\vartheta}\dot{x}\ddot{x} - i\varepsilon\dot{\bar{\vartheta}}\dot{x}^2 + i\varepsilon m^2\dot{\bar{\vartheta}} - 2i\varepsilon\dot{\vartheta}\bar{\vartheta}\dot{\bar{\vartheta}} + 2\vartheta i\varepsilon\bar{\vartheta}\ddot{\bar{\vartheta}}\\
= i\varepsilon\left\{(\dot{x}^2 + m^2)\bar{\vartheta} - 2\vartheta\dot{\bar{\vartheta}}\bar{\vartheta}\right\}^{\bullet} \quad .)
\end{aligned}$$

Now, however, the BRST variation δ is no longer nilpotent, but proportional to $\dot{x}^2 + m^2 + 2\vartheta\dot{\bar{\vartheta}}$. Since the action is invariant under the BRST variation by Noether's theorem there exists a conserved current. In order to derive a useful form for this current, we integrate the last term in the action by parts so that the action becomes

$$\int d\tau\mathcal{L},$$

with

$$\mathcal{L} = \dot{x}(\tau)^2 + m^2 + 2\dot{\vartheta}\bar{\vartheta}.$$

The Noether current then is

$$\begin{aligned}
&\frac{\delta\mathcal{L}}{\delta\dot{x}}\frac{\delta\dot{x}}{\delta\varepsilon} \quad + \quad \frac{\delta\mathcal{L}}{\delta\dot{\vartheta}}\frac{\delta\dot{\vartheta}}{\delta\varepsilon}\\
&= 2i\dot{x}^2\bar{\vartheta} + 2i(-\frac{1}{2}\dot{x}^2 + \frac{1}{2}m^2 + \vartheta\dot{\bar{\vartheta}})\bar{\vartheta}\\
&= i\bar{\vartheta}\left(\dot{x}^2 + m^2 + 2\vartheta\dot{\bar{\vartheta}}\right)\\
&=: Q_B.
\end{aligned}$$

Since the equations of motion for the fermionic variables are

$$\dot{\vartheta} = 0 = \dot{\bar{\vartheta}},$$

we are essentially left with

$$i\bar{\vartheta}(\dot{x}^2 + m^2).$$

If we consider the quantized variables, with

$$p = i\frac{\partial}{\partial x}$$

we have the commutation relations

$$[p^j, x^k]_- = -i\delta^{jk} \;, \quad [\vartheta, \bar{\vartheta}]_+ = 1$$

We denote the Hilbert space for the bosonic variables by H. It is infinite dimensional, whereas the Hilbert space for the fermionic variables is the finite dimensional space $\mathbb{C}^2$. Thus, the total Hilbert space is $H \otimes \mathbb{C}^2$. Following the physicists' notation, elements are written as

$$|k, \mu > .$$

We take a basis of eigenvectors $|k, \mu >, |k, \mu' >$ for the operators $p^j, \vartheta, \bar{\vartheta}$ satisfying

$$\begin{aligned} &p^j|k, \mu > = k^j|k, \mu >, \; p^j|k, \mu' > = \; k^j|k, \mu' > \\ &\vartheta|k, \mu > = 0, \; \vartheta|k, \mu' > = \; |k, \mu > \\ &\bar{\vartheta}|k, \mu > = |k, \mu' >, \; \bar{\vartheta}|k, \mu' > = 0. \end{aligned}$$

We then have

$$\begin{aligned} &Q_B|k, \mu > = -i(k^j k^j + m^2)|k, \mu' > \\ &Q_B|k, \mu' > = 0. \end{aligned}$$

We can now easily determine the BRST cohomology: The closed states are the $|k, \mu' >$ and those $|k, \mu >$ that satisfy $k^j k^j + m^2 = 0$, while the states $|k, \mu' >$ with $k^j k^j + m^2 \neq 0$ are exact. Thus, the states that are nontrivial in cohomology are the states $|k, \mu >$ and $|k, \mu' >$ with $k^j k^j + m^2 = 0$. This latter condition again is the bosonic part of the equations of motion. ([27])

Chapter 2

The Bosonic string

A reference for some of the sections of this chapter (more precisely, for parts of §§ 2.1, 2.4, 2.6, 2.7, 2.9, 2.10) is Albeverio, Jost, Paycha, Scarlatti [1]. A crucial difference to the presentation of [1], however, is that the present one avoids the use of harmonic maps which in contrast were constitutive for the approach to Teichmüller theory in [1].

2.1 The classical action for strings

We now consider a two-dimensional analogue of the point particle. We first investigate the Euclidean case because that one is more familiar than the Minkowski case. The analogue of a curve or path with given end points as studied in the previous chapter is a surface with given boundary. We thus consider two configurations Γ_1, Γ_2 of smooth closed oriented pairwise disjoint Jordan curves in $\mathbb{R}^d$, and a differentiable surface S with oriented boundary ∂S, equipped with a local parameter $z = (z^1, z^2)$, and a map

$$X : S \to \mathbb{R}^d$$

that maps ∂S diffeomorphically with preserved orientation onto $\Gamma = \Gamma_1 \cup \Gamma_2$. The area of the surface $X(S)$ is given by

$$A(X,S) = \int_S \sqrt{\det\left(\frac{\partial X^\mu}{\partial z^i}\frac{\partial X^\mu}{\partial z^j}\right)}\, dz^1\, dz^2. \tag{2.1.1}$$

In string theory, $A(X,S)$ is called the **Nambu-Goto action**.
With $\gamma_{ij} := \frac{\partial X^\mu}{\partial z^i}\frac{\partial X^\mu}{\partial z^j}$, $(\gamma^{ij}) := (\gamma_{ij})^{-1}, \det\gamma = \gamma_{11}\gamma_{22} - \gamma_{12}^2$ the Euler-Lagrange equations for (2.1.1) become

$$\frac{1}{\sqrt{(\det\gamma)}}\frac{\partial}{\partial z^i}\left(\sqrt{\det\gamma}\,\gamma^{ij}\,\frac{\partial X^\mu}{\partial z^j}\right) = 0 \quad \text{for} \quad \mu = 1, ..., d. \tag{2.1.2}$$

(γ_{ij}) is the metric on the surface $X(S)$ induced by the ambient Euclidean metric of $\mathbb{R}^d$; thus, it depends on the mapping X, and so (2.1.2) as an equation for X is highly nonlinear. Since the metric is not intrinsically determined on S, (2.1.1), and hence also (2.1.2) is invariant under the full diffeomorphism group of S; for any diffeomorphism

$$\varphi : S \to S,$$

we have

$$A(X \circ \varphi, S) \ = \ A(X, S). \tag{2.1.3}$$

In order to resolve these difficulties, and to remove the unpleasant square root from our action functional, we follow the same strategy as in Chapter 1, namely introduce a metric (g_{ij}) on S as an additional variable, with $g^{ij} = (g_{ij})^{-1}$, $\det g = g_{11}g_{22} - g_{12}^2$ as above and consider the so-called Dirichlet integral

$$D(X, g) := \frac{1}{2} \int_S g^{ij} \frac{\partial X^\mu}{\partial z^i} \frac{\partial X^\mu}{\partial z^j} \sqrt{\det g}\, dz^1\, dz^2 \tag{2.1.4}$$

In string theory, this is called the **Polyakov action**. The Euler-Lagrange equations for $D(X, g)$ w.r.t variations of X are

$$\frac{1}{\sqrt{\det g}} \frac{\partial}{\partial z^i} \left(\sqrt{\det g}\, g^{ij} \frac{\partial X^\mu}{\partial z^j} \right) \ = \ 0 \quad \text{for} \quad \mu = 1, ..., d \tag{2.1.5}$$

These are formally the same as (2.1.2), but since g_{ij} in contrast to γ_{ij} does not depend on X, these are linear equations for X.
(2.1.5) must be supplemented by a boundary condition, and in order to identify this boundary condition, we consider variations δX of X that preserve the requirement that ∂S be mapped to Γ infinitesimally in the sense that for each $z \in \partial S$, $\delta X(z)$ is tangent to Γ. If $D(X, g)$ is stationary w.r.t all such variations, we get

$$\begin{aligned} 0 = \frac{d}{dt} D(X + t\delta X, g)_{|t=0} = \\ & - \int_S \frac{1}{\sqrt{\det g}} \frac{\partial}{\partial z^i} \left(\sqrt{\det g}\, g^{ij} \frac{\partial X^\mu}{\partial z^j} \right) \delta X^\mu \sqrt{\det g}\, dz^1\, dz^2 \\ & + \int_{\partial S} \frac{\partial}{\partial n} X^\mu\, \delta X^\mu\, d\sigma(n) \end{aligned}$$

for all such δX, where $\frac{\partial}{\partial n}$ denotes the derivative in the direction of the exterior unit normal vector w.r.t the metric g_{ij} of ∂S. This implies (2.1.5) together with the Plateau boundary condition that

$$\frac{\partial}{\partial n} X^\mu \quad \text{is orthogonal to} \quad \Gamma$$

along ∂S, or, equivalently, if $\frac{\partial}{\partial t}$ denotes a tangential derivative along ∂S,

$$\frac{\partial}{\partial n} X^\mu \cdot \frac{\partial}{\partial t} X^\mu = 0$$

Varying g_{ij} is equivalent to varying g^{ij}, but the latter is more convenient for deriving the Euler-Lagrange equations for $D(X,g)$ w.r.t variations of the metric g; namely using

$$\delta \det g = (\det g) g^{ij}\, \delta g_{ij} = -(\det g) g_{ij}\, \delta g^{ij} \tag{2.1.6}$$

for the variations of the determinant, we get

$$T_{ij} := \frac{\partial X^\mu}{\partial z^i}\frac{\partial X^\mu}{\partial z^j} - \frac{1}{2} g_{ij}\, g^{kl} \frac{\partial X^\mu}{\partial z^k}\frac{\partial X^\mu}{\partial z^l} = 0 \tag{2.1.7}$$

for the so-called **energy-momentum tensor**. (2.1.5) means that X is harmonic w.r.t the metric (g_{ij}), while (2.1.7) means that X is conformal:

$$\frac{1}{\left(\det\left(\frac{\partial X^\mu}{\partial z^k}\frac{\partial X^\mu}{\partial z^l}\right)\right)^{\frac{1}{2}}} \frac{\partial X^\mu}{\partial z^i}\frac{\partial X^\mu}{\partial z^j} = \frac{1}{(\det(g_{kl}))^{\frac{1}{2}}}\, g_{ij}, \tag{2.1.8}$$

i.e. that the metric g_{ij} on S and the induced metric on $X(S)$ are proportional. Precisely if (2.1.8) holds,

$$D(X,g) \;=\; A(X,S).$$

i.e. the Polyakov and the Nambu-Goto action are equal, but in general, we only have the inequality

$$D(X,g) \geq A(X,S). \tag{2.1.9}$$

It is important to determine the invariances of $D(X,g)$:

1. Isometries of $\mathbb{R}^d$ (Euclidean motions applied to X) (in the case of a Minkowski target space, we have Poincaré invariance, i.e. invariance under the action of the Poincaré group applied to X)

2. Diffeomorphisms φ of S:

$$D(X,g) \;=\; D(X\circ\varphi, \varphi^* g)$$

3. Conformal transformations, i.e. multiplying (g_{ij}) pointwise by a conformal factor $\lambda(z) > 0$:

$$D(X,g) \;=\; D(X,\lambda g)$$

This **conformal**, or as it is also called in string theory, **Weyl invariance** is peculiar to two dimensions, as only here $\det g$ and g^{ij} have opposite conformal weights. Conformal invariance implies that the energy-momentum tensor is always traceless:

$$g^{ij}\, T_{ij} \;=\; T^i_i \;=\; 0. \tag{2.1.10}$$

Expressed in still another way, it means that the metric g_{ij} is only determined up to a conformal factor; thus, locally, we can choose a conformal gauge, i.e. assume that the metric is Euclidean, namely, we can introduce so-called conformal or isothermal coordinates so that the metric is of the form

$$g_{ij}(z) = \lambda(z)\,\delta_{ij}, \text{ for some positive function } \lambda(z) \tag{2.1.11}$$

by a theorem of Gauss, and then omit the conformal factor $\lambda(z)$ because of the conformal invariance of $D(X, g)$. We interpret this in the following way: We consider a Riemann surface Σ that is diffeomorphic to S, with a local **conformal** parameter $z = z^1 + iz^2 = u + iv$, and

$$D(X, \Sigma) = \frac{1}{2} \int_{\Sigma} \frac{\partial X^\mu}{\partial z^i} \frac{\partial X^\mu}{\partial z^i}\, dz^1\, dz^2. \tag{2.1.12}$$

If we write the coordinates as

$$z = z^1 + iz^2,$$

then the above form $\lambda(z)\delta_{ij}$ of the metric is preserved precisely under conformal (i.e. holomorphic, or else also antiholomorphic, but for the sake of simplicity, we ignore the second possibility) parameter transformations, i.e.

$$z = z(w) \quad \text{with} \quad \frac{\partial}{\partial \bar{w}} z(w) = 0.$$

Thus, in modern terminology, we may introduce a conformal structure Σ on S and then the metric becomes locally conformally Euclidean. Since our former variational problem

$$D(X, g) \to \min$$

determined the metric only up to a conformal factor, we have now found a way to remove that redundancy as we specify only a conformal structure and not a full metric anymore. $D(X, \Sigma)$ now is invariant only under **conformal** diffeomorphisms, but not under arbitrary ones, as the metric is locally fixed. We wish to study again the Euler-Lagrange equations for $D(X, \Sigma)$. Those equations w.r.t. variations of X are

$$\Delta X^\mu = 0 \quad \text{for} \quad \mu = 1, ..., d. \tag{2.1.13}$$

We put $u = z^1, v = z^2$, $X_u^2 = X_u^\mu X_u^\mu$ (Euclidean product) etc. and observe that (2.1.13) implies

$$\frac{\partial}{\partial \bar{z}}(X_z^2) = \frac{1}{2}\left(\frac{\partial}{\partial u} + i\frac{\partial}{\partial v}\right)(X_u - iX_v)^2 = 0.$$

Expressed in a more invariant manner, this means that

$$\phi\, dz^2 := 4X_z^2\, dz^2 \qquad (dz = du + idv) \tag{2.1.14}$$

is a holomorphic quadratic differential.
This object also arises if we consider variations of $D(X,\Sigma)$ through diffeomorphisms:
Let

$$\varphi_t : \Sigma \to \Sigma$$

be a smooth family of diffeomorphisms, with $\varphi_0 =$ identity.
We put

$$\varphi_t = \xi + i\eta, \qquad \frac{d\varphi_t}{dt}|_{t=0} = \nu + i\omega$$

Then

$$\begin{aligned} D(X \circ \varphi_t^{-1},\Sigma) & \\ &= \frac{1}{2}\int_\Sigma \left\{(X_u u_\xi + X_v v_\xi)^2 + (X_u u_\eta + X_v v_\eta)^2\right\} d\xi\, d\eta \\ &= \frac{1}{2}\int_\Sigma \left\{(X_u u_\xi + X_v v_\xi)^2 + (X_u u_\eta + X_v v_\eta)^2\right\} (\xi_u\eta_v - \xi_v\eta_u) du\, dv \end{aligned}$$

For $t = 0$, we have $u_\xi = 1 = v_\eta$, $u_\eta = 0 = v_\xi$ (because $\xi + i\eta = u + iv$ for $t = 0$). Thus, if we denote a derivative w.r.t t at $t = 0$ by a dot $^\bullet$,

$$\begin{aligned} \frac{d}{dt} D(X \circ \varphi_t^{-1},\Sigma)_{|t=0} &= \frac{1}{2}\int_\Sigma (2X_u^2 \dot{u}_\xi + 2X_v^2 \dot{v}_\eta + 2X_u\, X_v\, (\dot{v}_\xi + \dot{u}_\eta) \\ &\quad - (X_u^2 + X_v^2)\, (\dot{u}_\xi + \dot{v}_\eta)) du\, dv \\ &= -\frac{1}{2}\int \left\{(X_u^2 - X_v^2)(\nu_v - \omega_u) + 2X_u X_v(\nu_v + \omega_u)\right\} du\, dv \end{aligned}$$

since at $t = 0$, $\dot{u}_\xi = -\dot{\xi}_u = -\nu_u$ etc.

$$= \operatorname{Re} \int_\Sigma \phi\, (\nu + i\omega)_{\bar{z}}\, du\, dz \tag{2.1.15}$$

with

$$\phi := X_u^2 - X_v^2 - 2iX_u X_v.$$

Thus, if X is a critical point for $D(X,\Sigma)$ w.r.t variations by families of diffeomorphisms, we already obtain that (2.1.14) is a holomorphic quadratic differential.

If we choose the coordinates u and v near $\partial\Sigma$ in such a manner that u is tangential to $\partial\Sigma$, and if ϕ is continuous on $\partial\Sigma$, then we may use $\phi_{\bar{z}} = 0$ and integrate by parts to get

$$\operatorname{Re} \int_\Sigma \phi\, (\nu + i\omega)_{\bar{z}}\, du\, dv = \operatorname{Im} \int_{\partial\Sigma} \phi\, (\nu + i\omega)\, du \tag{2.1.16}$$

(The regularity question at the boundary is handled by Hildebrandt's theorem and its extensions, see § 2.3.)
We now assume that X is a critical point of $D(X, \Sigma)$ with respect to variations by all smooth families of diffeomorphisms

$$\varphi_t : \Sigma \to \Sigma;$$

the crucial point is that we do not require that the φ_t leave $\partial\Sigma$ pointwise fixed. This means that we do not impose a Dirichlet boundary condition for X, but only a weaker Plateau boundary condition, namely that X maps $\partial\Sigma$ onto the configuration Γ, without prescribing which point on $\partial\Sigma$ corresponds to a given point on Γ. That means that we may choose the real part ν of our variation arbitrarily, and from (2.1.15), (2.1.16), we see that the relation

$$\frac{d}{dt} D\left(X \circ \varphi_t^{-1}, \Sigma\right) = 0 \tag{2.1.17}$$

for all such variations implies

$$\operatorname{Im}\left(\phi \, dz^2_{|\partial\Sigma}\right) = 4X_u \cdot X_v \, du^2 = 0 \quad \text{on} \quad \partial\Sigma, \tag{2.1.18}$$

i.e. the holomorphic quadratic differential $\phi \, dz^2$ is real on $\partial\Sigma$.
Becoming bolder, we may ask for what families of diffeomorphisms (2.1.17) implies that $\phi \, dz^2$ itself vanishes; namely

$$\begin{aligned} \phi \, dz^2 = 0 &\Leftrightarrow X_u^2 = X_v^2 \quad \text{and} \quad X_u \cdot X_v = 0 \\ &\Leftrightarrow X \quad \text{maps} \quad \Sigma \quad \text{conformally onto} \quad X(\Sigma) \subset \mathbb{R}^d \end{aligned}$$

In other words, we are asking whether we can achieve (2.1.7), (2.1.8), the criticality of $D(X, g)$ w.r.t variations of the metric g, through the present variations by the group of diffeomorphisms. Since the Dirichlet integral is conformally invariant anyway, one might speculate that all the necessary variations come indeed from the diffeomorphism group. This is not quite the case, however, as we shall investigate in much detail below; at the moment, we can only take the following glimpse:
On a Riemann surface Σ, we may not necessarily be able to introduce a Euclidean metric globally; if we have some conformal metric $\lambda^2(z)dzd\bar{z}$, which is always possible, we can introduce an invariant L^2-product on the space of quadratic differentials:

$$(\phi \, dz^2, \psi dz^2) \equiv \frac{1}{2i} \int \phi\,(z)\bar{\psi}(z)\frac{1}{\lambda^2(z)} \, dz \, d\bar{z}; \tag{2.1.19}$$

"invariant" here means that it is invariant under conformal coordinate transformations.
Returning to (2.1.15), we thus see (2.1.17) would imply $\phi = 0$ if we can choose

$$(\nu + i\omega)_{\bar{z}} = \frac{1}{\lambda^2} \, \bar{\phi}. \tag{2.1.20}$$

Since ϕ is holomorphic, we have $\bar{\phi}_z = 0$, and so (2.1.20) implies

$$\frac{\partial}{\partial z}\left(\lambda^2(\nu + i\omega)_{\bar{z}}\right) = 0. \tag{2.1.21}$$

In the terminology of conformal analysis, this means that $\nu + i\omega$ is a so-called harmonic Beltrami differential. We shall see below that such a differential represents an infinitesimal variation of the complex structure of Σ. In particular, it cannot arise as an infinitesimal diffeomorphism of Σ. In fact, holomorphic quadratic differentials turn out to be orthogonal to all such infinitesimal diffeomorphisms, i.e. if as above $\nu + i\omega = \frac{d}{dt}\varphi_{t_{|t=0}}$ for a family of diffeomorphisms $\varphi_t : \Sigma \to \Sigma$, then automatically any holomorphic quadratic differential $\phi\; dz^2$ satisfies

$$\int \phi\, (\nu + i\omega)_{\bar{z}}\; du\; dv \;=\; 0,$$

so that this does not yield any restriction on ϕ.
In other words, the holomorphic quadratic differentials (or their dual objects, the harmonic Beltrami differentials, a point to be clarified below) represent precisely those variations that are orthogonal to those induced by the diffeomorphisms of the surface.
However, if Σ happens to be the unit disk

$$D \;=\; \{\, z \in \mathbb{C} \;:\; |z| < 1 \,\}$$

then any holomorphic quadratic differential $\phi\; dz^2$ that is real on ∂D already vanishes identically. Thus, in that particular case, any critical point of $D(X, \Sigma)$ w.r.t diffeomorphisms is conformal. This is of course related to the fact that by the Riemann mapping theorem, the conformal structure of the unit disc D is rigid and cannot be varied.

In the preceding, we have assumed that the boundary curves Γ are closed. If they are not closed, the surface S must have additional boundary pieces. For example, if Γ consists of two arcs, S might be a rectangle

$$S \;=\; \{0 \le u \le \ell\,,\; -T \le v \le T\} \tag{2.1.22}$$

and we require that $[0, l] \times \{-T\}$ is mapped to Γ_1 and $[0, l] \times \{T\} \to \Gamma_2$.
In that case, if we consider a variation

$$X + t\delta X$$

and compute the variation of (2.1.4)

$$\begin{aligned}\delta D(X, g) =& \frac{d}{dt} D(X + t\delta X, g)_{|t=0}\\ =& -\int_0^\ell \int_{-T}^{T} \frac{1}{\sqrt{\det g}} \frac{\partial}{\partial z^j}\left(\sqrt{\det g}\; g^{ij} \frac{\partial X^\mu}{\partial z^i}\right) \partial X^\mu \sqrt{\det g}\; dz^1\, dz^2 \qquad (2.1.23)\\ &- \int_{-T}^{T} \partial^u X^\mu \delta X^\mu \sqrt{\det g}\; dv\Big|_{u=0}^{u=\ell}\end{aligned}$$

(with $\partial^{z^i} = g^{ij}\partial_{z^j}$).
Since we may allow arbitrary variations δX^μ on $(\{u=0\}\cup\{u=\ell\})\times(-T,T)$,

$$\delta D(X,g) = 0$$

requires an additional boundary condition, like the Neumann condition

$$\partial^u X(0,v) = 0 = \partial^u X(\ell,v), \tag{2.1.24}$$

i.e. the normal derivative vanishes along $\{0\}\times(-T,T)$ and $\{\ell\}\times(-T,T)$. Another possibility would be the periodic boundary condition

$$\begin{aligned} X(0,v) &= X(\ell,v)\\ \partial^u X(0,v) &= \partial^u X(\ell,v)\\ g_{ij}(0,v) &= g_{ij}(\ell,v). \end{aligned}$$

This, however, means that the string is closed, and so we are returning to the case of closed boundary conditions.
It is important to observe that $D(X,g)$ is not the most general action exhibiting Poincaré, diffeomorphism, and conformal invariance. From general principles of quantum theory one requires that the action be polynomial in the field derivatives. The conformal invariance requires that the determinant $\sqrt{\det g}$ be compensated by a term of the form g^{ij}. In order to get a scalar Lagrange density, we need to contract each upper index with a derivative, and so each term in the density needs to contain two derivatives. One then checks that the requirements of diffeomorphism and Poincaré invariance only allow an additional term of the form

$$\frac{1}{2\pi}\int\limits_S K_g\sqrt{\det g}\,dz^1\,dz^2$$

for a closed surface, resp.

$$\frac{1}{2\pi}\int\limits_S K_g\sqrt{\det g}\,dz^1\,dz^2 \quad + \frac{1}{2\pi}\int\limits_{\partial S} k_g d\sigma(s),$$

for a surface with boundary, where K_g is the Gauss curvature of the metric g, and k_g is the geodesic curvature of ∂S w.r.t g.
These expressions are normalized, so that by the Gauss-Bonnet theorem, they equal the Euler characteristic $\chi(S)$. Thus, we may add to $D(X,g)$ a multiple of $\chi(S)$, i.e. consider

$$\frac{1}{2\pi\alpha'}D(X,g) + \lambda\chi(S),$$

with coupling constants $\frac{1}{4\pi\alpha'}$ and λ. α' is called Regge slope, while λ is the so-called string coupling constant.
As long as the genus of S is fixed, the term $\lambda\chi(S)$ is of no significance as it is a topological quantity independent of X and g. If, however, we consider the

partition function

$$\int dX\, dg\, e^{-\frac{1}{2\pi\alpha'}D(X,g)-\lambda\chi(S)}$$

and integrate not only over all metrics on a fixed topological surface, but also over all possible topological types, then λ determines the relative weights of the contributions of the various types.

Next, one may replace the flat target space (Euclidean or Minkowski) by a possibly curved one, i.e. introduce a (Riemannian or Lorentzian) metric $G_{\mu\nu}$ and consider

$$\frac{1}{4\pi\alpha'}\int\limits_S G_{\mu\nu}(X)g^{ij}\,\frac{\partial X^\mu}{\partial z^i}\,\frac{\partial X^\nu}{\partial z^j}\,\sqrt{\det g}\,dz^1dz^2.$$

If $G_{\mu\nu}$ transforms as a tensor, this action again is invariant under coordinate transformations on the target space, which we now denote by N. This action in fact is the one defining the nonlinear sigma model for the, say, Riemannian manifold $(N, G_{\mu\nu})$, or, in mathematical terminology, for harmonic maps from S to $(N, G_{\mu\nu})$. The Euler-Lagrange equations for this action are

$$\frac{1}{\sqrt{\det g}}\,\frac{\partial}{\partial z^i}\left(\sqrt{\det g}\,g^{ij}\frac{\partial X^\mu}{\partial z^j}\right)\,+\,g^{ij}\Gamma^\mu_{\nu\rho}\,\frac{\partial X^\nu}{\partial z^i}\,\frac{\partial X^\rho}{\partial z^j}=0\quad\text{for}\quad\mu=1,...,d,$$

with the so-called Christoffel symbols

$$\Gamma^\mu_{\nu\rho}:=\frac{1}{2}G^{\mu\sigma}(G_{\sigma\nu,\rho}+G_{\sigma\rho,\nu}-G_{\nu\rho,\sigma}),$$

where

$$G_{\sigma\nu,\rho}:=\frac{\partial}{\partial X^\rho}G_{\sigma\nu}.$$

Finally, one may add a term

$$\frac{1}{4\pi\alpha}\int B_{\mu\nu}(X)\varepsilon^{ij}\frac{\partial X^\mu}{\partial z^i}\,\frac{\partial X^\nu}{\partial z^j}\,\sqrt{\det g}\,dz^1\,dz^2,$$

where $B_{\mu\nu}(X)$ is an antisymmetric tensor, and ε^{ij} is determined by

$$\varepsilon^{ij}=-\varepsilon^{ji}\quad\text{for all}\quad i,j$$
$$\varepsilon^{12}=1.$$

Again, this is invariant under coordinate transformations in N, and, in addition, the integral is also invariant under variations

$$\delta B_{\mu\nu}(X)=\partial_\mu\xi_\nu(X)-\partial_\nu\xi_\mu(X).$$

(The 3-index field strength

$$H_{\mu\nu\rho}:=\partial_\mu B_{\nu\rho}+\partial_\nu B_{\rho\mu}+\partial_\rho B_{\mu\nu}$$

is pointwise invariant).

2.2 Sobolev spaces

In these notes, we shall frequently employ Sobolev spaces, and it is the aim of the present section to summarize the basic results about Sobolev spaces. We shall omit all proofs as these can be readily found e.g. in [18] or [32]. These spaces represent generalizations of the classical Lebesgue spaces (L^2, and more generally, L^p, for $1 \leq p \leq \infty$), and like the latter, their essential feature is that they are complete, i.e. Hilbert or Banach spaces, w.r.t certain integral norms. In distinction to the Lebesgue spaces, here those integral norms also involve derivatives.

We start with the Hilbert space case as that one will be the one figuring most prominently in our present treatise. We consider a bounded domain Ω in $\mathbb{R}^m$ with a smooth boundary. The restriction to domains in Euclidean space is not a serious one as all constructions can readily be generalized to Riemannian manifolds with the help of local coordinates. The requirement that the boundary is smooth avoids certain technical complications that will not be relevant for our purposes. (In our applications, we shall encounter a Riemann surface Σ, mostly with nonempty (and smooth) boundary, in place of the domain Ω, but, as said, this will not make much of a difference.)

We then define the Sobolev space $H^s(\Omega)(s \in \mathbb{N})$ simply as the completion of $C^\infty(\Omega)$, or, equivalently, of $C^s(\Omega)$ w.r.t the norm defined through

$$(2.2.1) \qquad ||f||^2_{H^s} := \sum_{\substack{i_1,\ldots,i_m=0 \\ i_1+\ldots+i_m\leq s}}^{s} \int \left(\frac{\partial^{i_1+\ldots+i_m}}{(\partial u^1)^{i_1}\ldots(\partial u^m)^{i_m}} f \right)^2 du^1 \ldots du^m,$$

for $f : \Omega \to \mathbb{R}$.

Here, if $i_j = 0$, no derivative is taken at all, and so the term with $i_1 = \ldots = i_m = 0$ simply yields the square of the standard L^2-norm. The squared H^s-norm of f then is the sum of the L^2-norms of all derivatives of f up to (and including) order s.

Certain facts about the spaces H^s then are obvious. They are complete and in fact Hilbert spaces, with the obvious definition of the scalar product. Moreover, the definition is easily adapted to the case of complex valued functions.

Furthermore, Sobolev functions can be integrated by parts. Namely each $f \in H^s$ possesses so-called weak derivatives $D_{i_1,\ldots,i_m} f \quad (i_1 + \ldots + i_m \leq s)$ that play the rôle of the classical partial derivatives $\frac{\partial^{i_1+\ldots+i_m}}{(\partial u^1)^{i_1}\ldots(\partial u^m)^{i_m}} f$ in the smooth case w.r.t integration, and coincide with them for smooth f.

Thus, for $f, g \in H^s$

$$(2.2.2) \qquad \int_\Omega f \cdot D_{i_1,\ldots,i_m} g = (-1)^{i_1+\ldots+i_m} \int_\Omega D_{i_1,\ldots,i_m} f \cdot g$$

These weak derivatives are given by L^2-functions, but in general they need not be smooth or even bounded. However, the remarkable embedding theorem

of Sobolev states that if a function is contained in the Sobolev space H^s for sufficiently high s, then it is actually continuous and differentiable up to a certain order k depending on s. We now proceed to state this result in more precise terms, restricting ourselves to the case $m = 2$ as being the one of interest for us. First of all, however, let us display an example of a function $f \in H^1(\Omega)$, here $\Omega \subset\subset D = \{z \in \mathbb{C} : |z| < 1\}$ the unit disc, that is not continuous, in fact not even bounded.
The example is

$$f(z) = \log(-\log r) \quad , \text{ where } z = re^{i\vartheta} \text{ (polar coordinates).}$$

We have

$$\begin{aligned} ||f||^2_{H^1(\Omega)} &= \int_\Omega (\log(-\log r))^2 r dr d\vartheta \quad + \int_\Omega \frac{1}{r^2(\log r)^2} r dr d\vartheta \\ &< \infty. \end{aligned}$$

(There are some easy technical points to check here, namely that f can be approximated by smooth functions in the Sobolev norm; in fact, a general theorem states that this holds true already if f possesses weak derivatives of the appropriate order satisfying the integration by parts formula for all smooth g, and so in the present case, one only needs to check that formula for the weak derivative w.r.t r, namely $\frac{-1}{r \log r} \in L^2(\Omega)$.)
The Sobolev embedding theorem in dimension 2 then states that, for any bounded domain Ω in $\mathbb{R}^2$ with smooth boundary,

$$H^s(\Omega) \subset C^k(\Omega) \quad \text{for } 0 \leq k < s - 1 \ (k \in \mathbb{N} \cup \{0\}).$$

As an easy extension of the preceding example shows, the result is sharp in the sense that it does not continue to hold for $k = s - 1$. However, a more precise result of Morrey improves the preceeding result to

$$H^s(\Omega) \subset C^{k,\alpha}(\Omega) \quad \text{for } 0 \leq k + \alpha < s - 1$$

$(k \in \mathbb{N} \cup \{0\}, 0 < \alpha < 1))$. $C^{k,\alpha}(\Omega)$ here is the space of functions on Ω that are k times continuously differentiable and whose k^{th}derivatives are Hölder continuous with exponent α.

An important consequence of Sobolev's embedding theorem is the result that if a function f is of class $H^s(\Omega)$ for all $s \in \mathbb{N}$, then it is also of class $C^\infty(\Omega)$, i. e. infinitely often differentiable. This provides the key to many regularity proofs for solutions of partial differential equations, as often Sobolev spaces are much better adapted to the setting of regularity proofs, in particular if the equations arise as Euler-Lagrange equations of variational problems, than the spaces $C^k(\Omega)$ of differentiable functions.
The result just mentioned can also be expressed as

$$C^\infty(\Omega) = \bigcap_{s \in \mathbb{N}} H^s(\Omega).$$

In particular, it implies that the Fréchet space $C^\infty(\Omega)$ is an inverse limit of Hilbert spaces. This often allows to overcome some of the limitations that Fréchet spaces exhibit when compared with Banach or Hilbert spaces, e.g. difficulties with implicit function theorems.
While for almost all of the present treatise, the Sobolev spaces $H^s(\Sigma)$ suffice, at certain places in the next section, we shall utilize a slightly more general class of spaces, namely the spaces $H^{s,p}(\Omega)$ (the space $H^{s,2}(\Omega)$ is our space $H^s(\Omega)$ just considered, but we now need another index to indicate the power to which the functions are integrable). These spaces are defined in an entirely analogous manner as before as those spaces that possess (weak) derivatives of order up to s that are integrable to the pth power ($1 \leq p < \infty$). The spaces $H^{s,p}(\Omega)$ then are again complete by definition and thus turn out to be Banach spaces.

One useful technical result about these spaces is the trace theorem. If we have an L^p function on Ω, its values on the lower-dimensional set $\partial\Omega$ are undefined, because L^p functions may have too large sets of discontinuity.
However, as we have seen from the Sobolev embedding theorem, if we also require integrability conditions for certain derivatives, i.e. consider Sobolev functions, the local control gets improved. The trace theorem then states that a function f of class $H^{s,p}(\Omega)$ admits a well-defined trace $f_{|\partial\Omega}$ of class $H^{s-1,p}(\partial\Omega)$, or, for that matter, such a trace $f_{|M}$ on any smooth $(m-1)$-dimensional submanifold M of Ω. "Well-defined" here means that the restriction $f_{|\partial\Omega}$ satisfies those properties that one would naturally require in the present context.

2.3 Boundary regularity

In this section, we study regularity questions for critical points of the Polyakov action. While these results are needed to make the approach presented here analytically rigorous, a reader more interested in the global aspects might wish to skip this section and feel assured that the analytic treatment presents no problems.
Interior regularity in fact is very easy. The real issue arises at the boundary, and as we have decided to consider strings with boundary, we cannot circumvent this issue.
In the classical treatments of minimal surfaces with boundary in the works of Douglas, Radó, and Courant, only continuity at the boundary was established, and in fact, below we shall display Courant's proof of continuity at the boundary. The breakthrough for more refined boundary regularity was only achieved much later by Hildebrandt, with subsequent extensions by many others. The generality needed for the present purposes has been realized in [17], and we shall essentially follow the reasoning given there.
We shall first investigate the regularity of the holomorphic quadratic differential associated with a map X from a Riemann surface Σ into Euclidean space or a Riemannian manifold that is stationary with respect to variations of the independent variables for the Polyakov action. As we had already observed, the

holomorphicity of that differential can also be derived from variations of the dependent variables. In any case, that regularity will rigorously establish that the holomorphic quadratic differential is real at the boundary of Σ. After that, we shall turn to the regularity of a map itself that is stationary with respect to the dependent variables.

Σ is a compact Riemann surface with boundary, and we consider the boundary $\partial\Sigma$ always as part of Σ. Thus, Σ is closed as a topological space. By uniformization, i.e. by choosing appropriate local conformal coordinates, we may always achieve a coordinate representation in which $\partial\Sigma$ is a (collection of) smooth curves. For example, we may assume that the boundary component of Σ under consideration is given by the unit circle in the plane.

Lemma 2.3.1. *Let Σ be a compact Riemann surface with a smooth boundary $\partial\Sigma$. Suppose $X : \Sigma \to \mathbb{R}^d$ satisfies*

$$\frac{d}{dt} D(X \circ \varphi_t^{-1}, \Sigma) = 0 \tag{2.3.1}$$

for all smooth families of diffeomorphisms

$$\varphi_t : \Sigma \to \Sigma \quad (-\varepsilon < t < \varepsilon \text{ for some } \varepsilon > 0)$$

with $\varphi_0 = id$. Then, with local conformal coordinates $z = u + iv$ on Σ,

$$\phi dz^2 = 4X_z^2 dz^2 = X_u^2 du^2 - X_v^2 dv^2 - 2iX_u \cdot X_v du dv \tag{2.3.2}$$

is a holomorphic quadratic differential that is real on $\partial\Sigma$, and in particular smooth up to the boundary of Σ.

Remarks:

1. A Riemann surface is a differentiable surface with a conformal structure. We implicitly assume that Σ does not have isolated boundary points. The smoothness of the boundary can then in fact be assumed or achieved by uniformization, i.e. by the appropriate choice of our local conformal coordinates $z = u+iv$. We do not want to explore this issue here, however, as the Riemann surfaces occuring in the sequel will always be given in suitable coordinates so that the boundary is smooth.

2. We have derived already in the preceding section that ϕdz^2 is a holomorphic quadratic differential, and that it is real on $\partial\Sigma$, provided a certain integration by parts is valid. The latter technical point is what the lemma is essentially about, and we shall show the regularity of ϕ up to the boundary precisely to clarify that issue.

3. The result and the proof to be given shortly continue to hold for the action for the nonlinear sigma model and even the general action

$$\frac{1}{4\pi\alpha'} \int G_{\mu\nu}(X)\, g^{ij}\, \frac{\partial X^\mu}{\partial z^i} \frac{\partial X^\nu}{\partial z^j} \sqrt{\det g}\, dz^1\, dz^2$$
$$+ \frac{1}{4\pi\alpha} \int B_{\mu\nu}(X)\, \varepsilon^{ij}\, \frac{\partial X^\mu}{\partial z^i} \frac{\partial X^\nu}{\partial z^j} \sqrt{\det g}\, dz^1\, dz^2$$

with symmetric $G_{\mu\nu}$ and antisymmetric $B_{\mu\nu}$ as discussed at the end of § 2.1.

The corresponding holomorphic quadratic differential then is

$$\phi(z)dz^2 = G_{\mu\nu}(X)(X_u^\mu X_u^\nu du^2 - X_v^\mu X_v^\nu dv^2 - 2iX_u^\mu X_v^\nu du\, dv)$$

(note that $B_{\mu\nu}$ does not enter here; the corresponding terms cancel because of the skew symmetry).

Proof. Let γ be a component of $\partial\Sigma$. We choose our local coordinates $z = u + iv$ in a neighborhood of γ in such a manner that this neighborhood is represented as an annulus A in the plane, with outer boundary $\{u + iv \in \mathbb{C}, u^2 + v^2 = 1\}$ corresponding to γ. We shall also utilize polar coordinates

$$r\, e^{i\vartheta} = u + iv.$$

As before, we put

$$\varphi_t = \xi + i\eta, \qquad \frac{d\varphi_t}{dt}_{|t=0} = \nu + i\omega.$$

For our purposes, we now let

$$\nu + i\omega = \Lambda(\vartheta)e^{i\vartheta} \tag{2.3.3}$$

near $\{r = 1\}$ and let it vanish near the other boundary component of A. From our assumption (2.3.1) and the computation in § 2.1 (see (2.1.15)), we obtain

$$\mathrm{Re} \int \phi(\nu + i\omega)_{\bar{z}}\, dudv = 0.$$

Standard potential theory then implies that ϕ is smooth, since weakly holomorphic, in the interior of A. We do not know yet, however, whether it is also smooth at the boundary $\{r = 1\}$. Therefore, at this moment, we may only consider $\gamma_\rho := \{r = \rho\}$ for $\rho < 1$ and integrate by parts to conclude

$$\lim_{\rho\to 1} \mathrm{Im} \int_{\gamma_\rho} \phi(\nu + i\omega)ie^{i\vartheta} d\vartheta = 0 \tag{2.3.4}$$

We select $a, b \in \mathbb{C}$ to kill the periods of ϕ and $z\phi$ along γ_ρ for $\rho < 1$, but close to one, i. e.

$$\int_{\gamma_\rho} (\phi + \frac{a}{z} + \frac{b}{z^2})\, dz = 0 \tag{2.3.5}$$

$$\int_{\gamma_\rho} z\, (\phi + \frac{a}{z} + \frac{b}{z^2})\, dz = 0. \tag{2.3.6}$$

We may then find an analytic function ψ on A with

$$\psi''(z) = \phi(z) + \frac{a}{z} + \frac{b}{z^2}. \tag{2.3.7}$$

Since X is of Sobolev class $H^1(\Sigma)$, ϕ is contained in the Lebesgue space $L^1(A)$, and ψ is then continuous on the closed annulus A, in particular on $\{r = 1\}$. Using (2.3.3) in (2.3.4), we obtain with (2.3.7)

$$0 = \lim_{\rho\to 1} \int_{\gamma_\rho} \Lambda(\vartheta)(\mathrm{Im}(\psi''(re^{i\vartheta})e^{2i\vartheta} + \mathrm{Im}(-ae^{i\vartheta}) + \mathrm{Im}(-b))d\vartheta$$

Since ϕ and hence ψ'' is in $L^1(A), \psi'$ has a trace of class L^1 on the boundary component γ of A. We therefore integrate the preceding formula by parts to convert the second derivative ψ'' into a first derivative ψ', and may subsequently put $\rho = 1$, in order to integrate over $\gamma = \gamma_1$.
This yields

$$\begin{aligned}
&-\mathrm{Im}\int_\gamma \frac{d}{d\vartheta}\Lambda(\vartheta) i\, e^{i\vartheta} - \mathrm{Im}\int_\gamma \frac{d}{d\vartheta}\Lambda(\vartheta) i\, b \log e^{i\vartheta}\, d\vartheta\\
&= \mathrm{Im}\int_\gamma \psi' i e^{i\vartheta}\frac{d}{d\vartheta}\Lambda(\vartheta)d\vartheta - \mathrm{Im}\int_\gamma \psi' e^{i\vartheta}\Lambda(\vartheta)d\vartheta\\
&= \mathrm{Im}\int_\gamma \psi' i e^{i\vartheta}\frac{d}{d\vartheta}\Lambda(\vartheta)d\vartheta - \mathrm{Im}\int_\gamma \psi' e^{i\tau}(\Lambda(0) + \int_0^\tau \frac{d}{d\vartheta}\Lambda(\vartheta)d\vartheta)d\tau\\
&= \mathrm{Im}\int_\gamma \frac{d}{d\vartheta}\Lambda(\vartheta)(ie^{i\vartheta}\psi'(e^{i\vartheta}) - i\psi(e^{i\vartheta}))d\vartheta
\end{aligned}$$

since the remaining term can be written as $\mathrm{Im}(i\Lambda(0)\int_0^{2\pi} \frac{d}{d\tau}\psi(e^{i\tau})d\tau)$ and therefore seen to vanish.
As $\Lambda(\vartheta)$ is arbitrary, we infer

$$\mathrm{Im}(i\, z\, \psi'(z) - i\psi(z) + i\, a\, z + i\, b\, \log z) = \text{ const on } \gamma. \tag{2.3.8}$$

Therefore, $i\, z\, \psi'(z) - i\, \psi(z) + ia\, z$ is analytic along γ, and thus ψ is smooth there. We differentiate (2.3.8) w.r.t. ϑ to obtain

$$\begin{aligned}
0 &= \mathrm{Im}(i\, z\frac{d}{dz}(i\, z\, \psi' - i\psi + i\, a\, z + i\, b \log z))\\
&= \mathrm{Im}(-z^2\phi(z)) \text{ on } \gamma.
\end{aligned}$$

Thus, the harmonic function $\mathrm{Im}(z^2\phi(z))$ has vanishing boundary values on γ, and it is therefore smooth in A by standard potential theory, up to the boundary γ. In particular, its derivative normal to γ is controlled. The Cauchy-Riemann equations for the holomorphic function $z^2\phi(z)$ then also control the real part, and so $z^2\phi(z)$ is smooth up to γ, and hence so is ϕ.
In particular, our integration by parts on γ is valid, and we see from (2.1.16) or (2.3.4) for $\rho = 1$ that $\phi\, d\, z^2$ is real at the boundary of Σ. □

Remark. The preceding argument is due to Morrey [24]. A different proof can be found in Courant's work, see [7].

We have thus established optimal regularity results for the holomorphic quadra tic differential $\phi\, d\, z^2$ defined by a map $X : \Sigma \to \mathbb{R}^d$ that is critical for the Polya kov action w.r.t variations of the independent variables. We now return to the regularity question for X itself. As mentioned in § 2.1, if X is critical for the Polyakov action w.r.t. variations of the dependent variables, it is harmonic,

$$\Delta X = 0.$$

Thus, as is well known, X is real analytic in the interior of Σ, and so interior regularity is not an issue. The question about the regularity of X at the boundary $\partial\Sigma$ is more subtle, and, naturally, it will depend on the boundary condition imposed. We have argued that the condition for X at the boundary should be of Plateau type, i.e. $X : \partial\Sigma \to \Gamma$ should be an oriented diffeomorphism, or, since that condition is not closed in the function space naturally adapted to the problem, namely the Sobolev space H^1, a limit of such diffeomorphisms. This is expressed by saying that X should map $\partial\Sigma$ monotonically onto Γ (and with preserved orientation on each component of $\partial\Sigma$). Our first step now is to show that any harmonic $X : \Sigma \to \mathbb{R}^d$ that maps $\partial\Sigma$ monotonically onto Γ in fact is continuous, even without requiring that X be stationary w.r.t. to all variations of the independent variables. For that purpose, we shall use a calculus lemma, the so-called Courant-Lebesgue lemma whose easy proof may for example be found in [17]:

Lemma 2.3.2. *Suppose that $p \in \partial\Sigma$, and that near p local conformal parameters $z = u + iv$ are chosen such that p corresponds to 0, some neighborhood of p in Σ to $B_1^+ := \{u + iv \in \mathbb{C} : u^2 + v^2 < 1, \quad v > 0\}$ (thus, $\partial\Sigma$ is represented near p as $\{v = 0,\ u^2 < 1\}$). We also use polar coordinates $z = re^{i\vartheta}$. Let $X \in H^1(B^+, \mathbb{R}^d)$, with $D(X, B^+) \le K$. Then for every $0 < \delta < 1$, there exists some $\rho \in (\delta, \sqrt{\delta})$ for which the restriction of X to the semicircle $\{r = \rho,\ v \ge 0\}$ is absolutely continuous and*

$$|X(w_1) - X(w_2)| \le (8\pi K)^{\frac{1}{2}}\ (\log\frac{1}{\delta})^{-\frac{1}{2}} \tag{2.3.9}$$

for all $w_1, w_2 \in \{r = \rho,\ v > 0\}$.

(Note that in particular, the right hand side of (2.3.9) goes to 0 as δ tends to 0.)

Lemma 2.3.3. *Let $X : \Sigma \to \mathbb{R}^d$ be harmonic, mapping $\partial\Sigma$ monotonically onto a configuration Γ of closed Jordan curves in $\mathbb{R}^d$. If the Dirichlet integral (Polyakov action) of X is finite, then X is continuous up to the boundary. In particular, every component of $\partial\Sigma$ is mapped either to a point or surjectively onto the corresponding curve of Γ.*

Proof. Let $p \in \partial\Sigma$, and choose local coordinates near p as in the statement of Lemma 2.3.2. Suppose the component of $\partial\Sigma$ containing p is mapped to the closed Jordan curve γ. Let the Dirichlet integral of X be bounded by K. Given $\eta > 0$, we choose $\delta > 0$ such that

$$(8\pi K)^{\frac{1}{2}} (\log \frac{1}{\delta})^{-\frac{1}{2}} < \eta. \tag{2.3.10}$$

We then choose ρ as in Lemma 2.3.2, i.e. such that (2.3.9) holds. We let w_1, w_2 be the two points where the semicircle $\{r = \rho,\ v \geq 0\}$ meets the real axis $\{v = 0\}$. Thus, w_1, w_2 represent points on $\partial\Sigma$, and they are thus mapped to γ. The images $X(w_1), X(w_2)$ divide γ into two subarcs γ', γ''.
Since γ is a closed Jordan curve, for every $\varepsilon > 0$, we may find $\eta > 0$ such that for any two points $q_1, q_2 \in \gamma$ with $|q_1 - q_2| < \eta$ for all points q on one of the two subarcs into which q_1 and q_2 divide γ, we have

$$|q - q_1| < \frac{\varepsilon}{2}. \tag{2.3.11}$$

We may also assume $\eta < \frac{\varepsilon}{2}$. Given $\varepsilon > 0$, we then choose $\eta > 0$ so as to satisfy (2.3.11), and $\delta > 0$ in turn so as to satisfy (2.3.10). With $q_i = X(w_i), i = 1, 2$, we let γ' be the subarc satisfying (2.3.11). We then have to consider two possibilities:

1. The line segment $\{v = 0, -\rho \leq u \leq \rho\}$, i.e. the subarc of $\partial\Sigma$ containing our p, is mapped to γ'. In that case, the boundary of the semidisc

   $$\{v \geq 0, u^2 + v^2 \leq \rho^2\}$$

   is mapped into some region of diameter at most ε. Namely, on the semicircle $\{\tau = \rho, v \geq 0\}$, this is guaranteed by (2.3.10) (noting $\eta < \frac{\varepsilon}{2}$), and on the line $\{v = 0, -\rho \leq u \leq \rho\}$ by (2.3.11).

   The maximum principle for harmonic functions then implies that the whole semidisc itself is mapped into such a region. As $\varepsilon > 0$ is arbitrary, this establishes continuity.

2. The line segment $\{v = 0, -\rho \leq u \leq \rho\}$ is mapped to the longer subarc γ''.

   The situation might either be reversed, i.e. we might get back to case 1., if we decrease ε below a certain threshold $\varepsilon_0 > 0$, and we shall be in case 1. for all $\varepsilon < \varepsilon_0$, because of the monotonicity, or case 2. will persist for all $\varepsilon > 0$. In that situation, while the segment of $\partial\Sigma$ enclosed by the points w_1, w_2 and containing p is mapped to the longer subarc γ'' of γ, the remainder of the same component of $\partial\Sigma$ is mapped to the shorter subarc γ'. We call that subarc of $\partial\Sigma$ the exterior arc of p; thus, the component of $\partial\Sigma$ containing p is divided into two subarcs, one containing p, the other one called the exterior arc of p. As ε, and with it δ tends to 0, that exterior subarc eventually contains every point of our component of $\partial\Sigma$, except p itself. By (2.3.10) and the properties of γ' again, the closed curve in Σ consisting of that subarc and the semicircle $\{r = \rho, v \geq 0\}$ is mapped to a region of diameter at most ε, which becomes a point in the limit $\varepsilon \to 0$.

In particular, the whole component of $\partial\Sigma$ containing p, with the possible exception of p itself, is mapped to a single point. We may then, however, remove the possible discontinuity of p by simply mapping p to the same point.

This completes the proof of the lemma.

□

Remark. The same proof works more generally for the functionals discussed at the end of § 2.1. What we need is the finiteness of the Dirichlet integral, plus the maximum principle. The maximum principle may not be valid globally for nonlinear sigma models, but it does hold locally, and that suffices. For details, see [17]

For more precise regularity results, we shall need the so-called Hartman-Wintner Lemma:
For $X : \Sigma \to \mathbb{R}^d$, we employ the following complex notation:

$$X_z := \frac{1}{2}(X_u - iX_v) : \Sigma \to \mathbb{C}^d,$$

and similarly for $X_{\bar{z}}, X_{z\bar{z}}$, where $z = u + iv$ are local conformal coordinates as before.

Lemma 2.3.4. *Suppose $X \in C^{1,1}(\Sigma, \mathbb{R}^d)$ satisfies almost everywhere*

$$|X_{z\bar{z}}| \leq c_0(|X_z| + |X|) \text{ for some constand } c_0. \tag{2.3.12}$$

If

$$X(z) = \circ(|z|^n) \tag{2.3.13}$$

for some $n \in \mathbb{N}$, then

$$\lim_{z \to 0} X_z z^{-n} \tag{2.3.14}$$

exists. If (2.3.13) holds for all $n \in \mathbb{N}$, then

$$X \equiv 0.$$

For a *proof*, see [17].

Theorem 2.3.1. *Let Γ be a collection of C^2 closed Jordan curves in $\mathbb{R}^d, X : \Sigma \to \mathbb{R}^d$ harmonic, continuous, mapping $\partial\Sigma$ to Γ, with holomorphic quadratic differential*

$$\phi\, dz^2 = 4X_z^2\, dz^2$$

that is real on $\partial\Sigma$.

Then $X \in C^{1,\alpha}(\Sigma, \mathbb{R}^d)$, for some $0 < \alpha < 1$, and at each $z_0 \in \Sigma$, we have the asymptotic expansion

$$X_z = a(z - z_0)^m + \circ(|z - z_0|^m) \tag{2.3.15}$$

for some $a \in \mathbb{C}^d \backslash \{0\}$ and some nonnegative integer $m \geq 1$. (If X is conformal, then $\phi\, dz^2 = 0$, hence $a^2 = 0$.) If $\Gamma \in C^{k,\alpha}$ for some $k \geq 2$ and some $0 < \alpha < 1$, then $X \in C^{k,\alpha}$ as well.

Proof. We represent the boundary curve of Σ containing a given point $p_0 \in \partial\Sigma$ as the unit circle in the plane. We apply a coordinate transformation in the image so that in some neighborhood of $X(p_0)$, Γ is represented as

$$\Gamma = \{X^1 = X^2 = ... = X^{d-1} = 0, |X^d| < 1\}.$$

In general, this is a nonlinear coordinate transformation, and so the Laplace equation $\Delta X = 0$ gets transformed into the harmonic map equation.

$$\Delta X^\alpha + \Gamma^\alpha_{\beta\gamma} X^\beta_{z^i} X^\gamma_{z^i} = 0, \text{ for } \alpha = 1, ..., d, \tag{2.3.16}$$

with standard summation conventions. Here, $\Gamma^\alpha_{\beta\gamma} = \frac{1}{2} g^{\alpha\delta}(g_{\beta\delta,\gamma} + g_{\gamma\delta,\beta} - g_{\beta\gamma,\delta})$ are the Christopffel symbols for the image metric $g_{\alpha\beta}$ in our new coordinates. (We observe that the whole proof will then work more generally for solutions of the nonlinear sigma model as we shall work on the basis of (2.3.16).) By Lemma 2.3.1,

$$\phi = 4 g_{\alpha\beta}\, u^\alpha_z\, u^\beta_z \tag{2.3.17}$$

is holomorphic and smooth up to the boundary and $z^2\phi(z)$ (noting that our boundary curve is represented as the unit circle) is real at the boundary. We rewrite (2.3.17) as

$$(X^d_z + \frac{g_{\mu d}}{g_{dd}} X^\mu_z)^2 = (\frac{g_{\mu d}}{g_{dd}} X^\mu_z)^2 - \frac{g_{\mu\nu}}{g_{dd}} X^\mu_z X^\nu_z + \frac{\phi}{g_{dd}} \tag{2.3.18}$$

where μ and ν are summed from 1 to $d-1$, here and in the sequel.
Since $\Gamma \in C^2$, we may assume

$$g_{\alpha\beta} \in C^1, g_{\alpha\beta}(X(p_0)) = \delta_{\alpha\beta}.$$

This and the continuity of X allow to derive the following estimate from (2.3.18), in some neighborhood of p_0

$$|\bigtriangledown X^d|^2 \leq c_1 \sum_{\mu=1}^{d-1} |\bigtriangledown X^\mu|^2 + c_2, \tag{2.3.19}$$

for constants c_1, c_2, the latter depending on ϕ. Inserting this into (2.3.16), we get for the vector $Y := (X^1, ..., X^{d-1})$,

$$|\Delta Y| \leq c_3 |\bigtriangledown Y|^2 + c_2$$
$$Y = 0 \text{ on the boundary.}$$

If we recall that we already know that Y is continuous, this implies, by standard elliptic estimates, that

$$Y \in C^{1,\beta} \text{ near } p_0, \text{ for some } 0 < \beta < 1.$$

(2.3.19) then implies that $\bigtriangledown X^d$ is bounded as well. We also compute from (2.3.18)

$$(izX_z^d + \frac{g_{\mu d}}{g_{dd}} izX_z^\mu)^2 = \frac{g_{\mu\nu}}{g_{dd}}(zX_z^\mu)(zX_z^\nu) - (\frac{g_{\mu d}}{g_{dd}} zX_z^\mu)^2 - \frac{z^2\phi}{g_{dd}}. \tag{2.3.20}$$

On the boundary (the unit circle), we have

$$zX_z^k = \frac{1}{2}(X_r^k - X_\vartheta^k)$$

in polar coordinates $z = re^{i\vartheta}$. Since $X_r^\mu = 0$ at the boundary for $\mu = 1, ..., d-1$, and since $z^2\phi$ is real there, the right hand side of (2.3.20) is real. This implies

$$\begin{aligned} 0 &\equiv \left(X_r^d + \frac{g_{\mu d}}{g_{dd}} X_r^\mu\right) \left(X_\vartheta^d + \frac{g_{\mu d}}{g_{dd}} X_\vartheta^\mu\right) \\ &= \left(X_r^d + \frac{g_{\mu d}}{g_{dd}} X_r^\mu\right) X_\vartheta^d \end{aligned} \tag{2.3.21}$$

at the boundary.

We now construct a new function $\tilde{X}$ at a neighborhood $U(p_0)$ of p_0 by locally reflecting X across the boundary. For that purpose, for $\mu = 1, ..., d-1$, we put

$$\tilde{X}^\mu(re^{i\vartheta}) := \begin{cases} X^\mu(re^{i\vartheta}) & r \le 1 \\ -X^\mu(r^{-1}e^{i\vartheta}) & r \ge 1 \end{cases}$$

and

$$\tilde{X}^d(re^{i\vartheta}) := \begin{cases} X^d(re^{i\vartheta}) & r \le 1 \\ 2X^d(e^{i\vartheta}) - X^d(r^{-1}e^{i\vartheta}) & r \ge 1. \end{cases}$$

$\tilde{X}$ then is of class C^1 in $U(p_0)$, and of class C^2 except at the boundary circle. Hence $\tilde{X} \in C^{1,1}(U(p_0))$.

By (2.3.16) and since $\tilde{X}$ is of class C^1, we also have

$$|\tilde{X}_{z\bar{z}}| \le c_0 |\tilde{X}_z|, \tag{2.3.22}$$

i.e. the assumption of Lemma 2.3.4 (in fact, since we do not have the term $|X|$ on the right hand side here, this simplifies the argument of Lemma 2.3.4 considerably). We thus get

$$\tilde{X}_z = a(z - p_0)^m + o(|z - p_0|^m),$$

for some $a \in \mathbb{C}^d \backslash \{0\}, m \in \mathbb{N}$. This in turn implies that X^d_ϑ either vanishes identically, which is a trivial case, or that this happens only at finitely many points. In the latter case, we conclude from (2.3.21)

$$X^d_r = -\frac{g_{\mu d}}{g_{dd}} X^\mu_r \quad \text{on the boundary.}$$

(2.3.16) yields

$$\Delta X^d = -\Gamma^d_{\alpha\beta} X^\alpha_{z^i} X^\beta_{z^i}. \tag{2.3.23}$$

The right hand side is bounded as we already know that the gradient of X is bounded. In fact, since we even know that $X \in C^{1,\beta}$, the right hand side is of class C^α, for some $0 < \alpha < 1$, and linear elliptic theory implies $X \in C^{2,\alpha}$. Higher regularity then follows from the standard boot strap argument. Namely, if the $\Gamma^\alpha_{\beta\gamma}$ are of class $C^{1,\gamma}$ (which can be achieved if $\Gamma \in C^{2,\gamma}$, for some $0 < \gamma < 1$), then, since $X \in C^{2,\alpha}$, the right hand side of (2.3.23) is of class $C^{1,\delta}$, for some $0 < \delta < 1$, hence $X \in C^{3,\delta}$ by linear elliptic theory, and in this manner we get as much regularity as Γ permits. □

Remark. As mentioned already in the course of the proof, the nonlinear coordinate transformation in the image requires that we work with the nonlinear harmonic map equation. This has the benefit that the proof directly carries over to the nonlinear sigma model, and in fact, it is easily seen to cover also the general action discussed at the end of § 2.1.

2.4 Spaces of mappings and metrics

As in § 2.1, S is a compact surface, with boundary ∂S, in case our boundary configuration Γ is nonempty. While our considerations in the present § easily extend to the more general functionals studied at the end of the previous §, for simplicity of notation, here we only consider the Dirichlet integral or Polyakov action $D(X, g)$. We fix some background metric g_0 on S, and some map

$$X_0 : S \to \mathbb{R}^d$$

mapping ∂S diffeomorphically and with the prescribed orientation onto Γ. If X_0 is harmonic w.r.t g, i. e.

$$\Delta_g X_0 = 0, \tag{2.4.1}$$

and if we split an arbitrary $X : S \to \mathbb{R}^d$ satisfying our boundary condition as

$$X = X_0 + \xi, \qquad \xi \in H^s_0(S, \mathbb{R}^d) \quad \text{(Sobolev space as in § 1.3)}, \tag{2.4.2}$$

we have from (2.4.1) and $\xi_{|\partial S} = 0$

$$D(X_0 + \xi, g) = D(X_0, g) + D(\xi, g).$$

As in § 1.3, we consider the affine spaces

$$F^s := X_0 + H_0^s(S, \mathbb{R}^d),$$

with Riemannian structure given by

$$(V, W)_g := \int_S V^\mu(z) W^\mu(z) \sqrt{\det g}\, dz^1\, dz^2. \tag{2.4.3}$$

The regularity aspects are the same as in § 1.3, and so we again denote the group of diffeomorphisms of our domain, which now is the surface S, of Sobolev class H^ℓ by $\mathcal{D}^\ell$. In analogy to Lemma 1.3.1, we have

Lemma 2.4.1. $(V, W)_g = (V \circ \varphi, W \circ \varphi)_{\varphi^* g}$, *for* $\varphi \in \mathcal{D}^\ell$ *with* $(\varphi^* g)_{ij}(z) = g_{k\ell}(\varphi(\tau)) \frac{\partial \varphi^k}{\partial z^i} \frac{\partial \varphi^\ell}{\partial z^j}$, *i.e. the Riemannian structure on* F^s *is invariant under the action of the diffeomorphism group.*

By way of contrast, however,

Lemma 2.4.2. *The Riemannian structure* $(.,.)_g$ *on* F^s *is not invariant under conformal changes of the metric* g.

Namely, a conformal change is represented by

$$g_{ij}(z) \to \lambda^2(z) g_{ij}(z),$$

and we get a factor $\lambda^2(z)$ in the integrand in (2.4.3). This is a new phenomenon that had no analogue for the point particle of Chapter 1, and it leads to a so-called conformal anomaly in our attempts to define the partition function

$$\int dX\, dg\, e^{-D(X,g)}.$$

Namely, our strategy should be to drop the integration over the gauge degrees of freedom at the expense of introducing a Faddeev-Popov determinant as we did in Chapter 1. Lemma 2.4.1 tells us that at least as far as the X integration is concerned (and as we shall soon see, the same holds for the g integration), there is no obstruction for dividing out the action of the diffeomorphism group, but Lemma 2.4.2 implies that this is not so for the conformal transformations. While the classical action $D(X, g)$ is conformally invariant, this is not so anymore for the formal measure dX. We get a so-called **conformal anomaly** in our attempts to quantize the classical action.

However, as we shall now see, there exists another conformal anomaly as the formal measure dg is not conformally invariant either.

For that purpose, we consider the space $\mathcal{M}^k$ of metrics on S of Sobolev class H^k, again in close analogy to § 1.3.

The tangent space $T_g\mathcal{M}^k$ is given by symmetric 2×2 tensors $h = (h_{ij})$ of class H^k. Each such h can be decomposed into its trace and tracefree parts:

$$h = \rho g + h'\ ,\ \rho : S \to \mathbb{R}, \tag{2.4.4}$$

$$h'_{ij} = \frac{1}{2}(\delta_i^k \delta_j^\ell + \delta_i^\ell \delta_j^k - g_{ij} g^{k\ell}) h_{kl}.$$

The decomposition (2.4.4) is orthogonal w.r.t the natural Riemannian structure on $T_g\mathcal{M}^k$:

$$((h_{ij}),(\ell_{ij}))_{g,\kappa} := \int (g^{ijkm} + \kappa g^{ij} g^{km} h_{ij} \ell_{km}) \sqrt{\det g}\, dz^1\, dz^2 \tag{2.4.5}$$

with $\kappa > 0$ and

$$g^{ijkm} := \frac{1}{2}(g^{ik}g^{jm} + g^{im}g^{jk} - g^{ij}g^{km}).$$

Our subsequent considerations will not depend on the value of κ, and so for simplicity, we put $\kappa = \frac{1}{2}$ so that (2.4.5) becomes

$$((h_{ij}),(\ell_{ij}))_g := \int_S g^{ij} g^{km} h_{ik} \ell_{jm} \sqrt{\det g}\, dz^1\, dz^2. \tag{2.4.6}$$

As before, we get

Lemma 2.4.3. *The Riemannian metric $(.,.)_g$ on $T_g\mathcal{M}^k$ is invariant under the action of the diffeomorphism group, but not under conformal transformations.*

2.5 The global structure of the spaces of metrics, complex structures, and diffeomorphisms on a surface

Let S be a compact surface of genus p with k boundary curves. We form the Schottky double S^d of S by identifying S with another copy S' of S with opposite orientation along their common boundary. S^d then is a compact oriented surface without boundary of genus

$$q = 2p + k - 1.$$

Thus, if S is the disk ($p = 0, k = 1$), S^d is the Riemann sphere, if S is a cylinder ($p = 0, k = 2$), S^d becomes a torus, and in all other cases, S^d has genus at least 2. S^d carries an involution

$$i : S^d \to S^d \quad (\text{i.e. } i^2 = id\ )$$

that interchanges S and S' and has the boundary of S as its fixed point set and is orientation-reversing.
If S carries a Riemannian metric for which ∂S is geodesic, so does S^d, and i then becomes an isometry. ∂S as the fixed point set of an isometry has to be geodesic. If S is a Riemann surface, i.e. carries a complex structure, so does S^d, and i becomes antiholomorphic.

We shall now need to discuss some global structures on the spaces of metrics, complex structures, diffeomorphisms etc. of a given closed surface. We shall adopt the approach of Tromba [30].

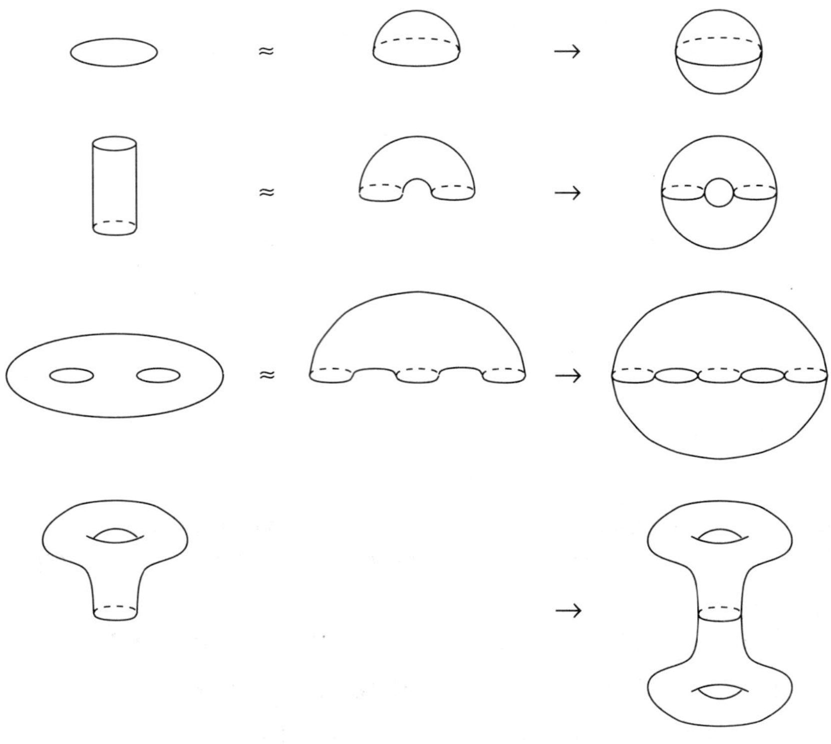

Figure 2.1: Going from S to its Schottky double S^d thus preserves all the important possible structures on S and reduces the study of surfaces with boundary to the one of closed surfaces.

As one knows, the cases $q = 0$, $q = 1$, and $q \geq 2$ correspond to different geometries. Namely, a closed surface of genus $q = 0$, i.e. a sphere, carries a metric of curvature 1, a closed surface of genus $q = 1$, i.e. a torus, one of vanishing curvature, and finally a closed surface of genus $q \geq 2$, a hyperbolic metric, i.e. one of curvature -1.
More precisely, by the Poincaré uniformization theorem, given a closed surface S (we write S in place of S^d for simplicity) equipped with a Riemannian metric (g_{ij}), there exists a conformally equivalent surface with constant curvature; in other words there exists a positive function $\rho : S \to \mathbb{R}^+$ such that the metric

$$\rho^2 g_{ij}$$

on S has constant curvature $1, 0$, or -1, resp. If in the case of vanishing curvature, we stipulate in addition that the area of this metric be normalized to 1, given (g_{ij}), the function ρ with this property is uniquely determined for curvature $c = 0, -1$. In fact, if we denote the space of metrics of Sobolev class k and constant curvature c by $\mathcal{M}_c^k(c = 1, 0, -1,$ resp.), then for every positive function $\lambda : S \to \mathbb{R}$ (of the appropriate regularity class H^k) and every metric $g \in \mathcal{M}_c^k$, the family of metrics $(1 + t\lambda)_g^2$, with the condition that the area is unchanged in case $c = 0$, is transversal to $\mathcal{M}_c^k$ at $t = 0$ for $c = 0, -1$. Of course, such a statement presupposes that the spaces $\mathcal{M}^k$ and $\mathcal{M}_c^k$ possess differentiable structures, a point to which we shall return shortly.

By the theorem of Gauss on the existence of local conformal parameters (Gauss proved this result in the smooth category; under more general regularity assumptions, e.g. for Sobolev metrics, this was investigated by Lichtenstein, Morrey, Ahlfors, Bers, and others, see [17]), given a metric (g_{ij}) on a surface S, we may always introduce local coordinates in which the metric takes the form

$$\sigma^2 \delta_{ij}$$

i.e. is conformally equivalent to the Euclidean one. If we identify the Euclidean space $\mathbb{R}^2$ in the standard manner with the complex plane $\mathbb{C}$, these local coordinates allow us to transport the local complex structures of the coordinate patch back to the surface S, as all oriented coordinate transitions between metrics of such a type preserve angles, hence are conformal, i.e. holomorphic. Thus, each surface S with a metric (g_{ij}) is equipped with a unique conformal or complex structure, i.e. becomes a Riemann surface.

We conclude that constant curvature metrics (with fixed area 1 in case $c = 0$) and complex structures on Riemann surfaces are in bijective correspondence.

In other words, if $\mathcal{C}^k$ denotes the space of positive functions of Sobolev class H^k on S, we obtain the identification

$$\begin{aligned} \mathcal{M}^k/_{\mathcal{C}^k} &\approx \mathcal{M}_c^k, \text{ for } c = -1 \text{ (genus } S > 1) \\ &\approx \bar{\mathcal{M}}_0^k = \{g \in \mathcal{M}_0^k : \text{Area } (S, g) = 1\} \quad \text{for} \quad c = 0 \quad (\text{genus } S = 1). \end{aligned}$$

At the moment, this is an identification between sets, but it turns out that both spaces possess differentiable structures so that the above identification becomes a diffeomorphism.

$\mathcal{M}^k$ is an open subset of the Hilbert space of all symmetric 2×2 tensors on S of Sobolev class H^k (namely the subset of everywhere positive definite tensors), and thus obviously becomes a Riemannian Hilbert manifold, as we have already utilized above.

In order to show that $\mathcal{M}_{-1}^k$ is a smooth manifold, one shows that -1 is a regular value for the scalar curvature function

$$R : \mathcal{M}^k \to H^{k-2}(S)$$

and then applies the implicit function theorem. In local conformal coordinates $z = x + iy$, i.e. if the metric g has the form

$$\sigma^2 \delta_{ij},$$

R is given by the formula

$$R = -2\Delta_g \log \sigma,$$

where $\Delta_g = \frac{4}{\sigma^2}\frac{\partial^2}{\partial z \partial \bar{z}} = \frac{1}{\sigma^2}(\frac{\partial^2}{\partial x^2} + \frac{\partial}{\partial y^2})$ is the Laplace-Beltrami operator for the metric $g = (\sigma^2 \delta_{ij})$. R is twice the usual Gauss curvature. In nonconformal coordinates, the formula for R becomes more complicated.

We now consider the action of the oriented diffeomorphism group $\mathcal{D}^{k+1}$ on the spaces $\mathcal{M}^k$ and $\mathcal{M}^k_c$. We shall also consider the subgroup $\mathcal{D}^\ell_0$ of $\mathcal{D}^\ell$ consisting of those diffeomorphisms that are homotopic to the identity of S. Since this will be helpful, for the moment we again allow S to possibly have a boundary. Of course, all diffeomorphisms then have to map ∂S to itself, and this property has to be preserved under all homotopies. By Baer's theorem, homotopic diffeomorphisms between surfaces are isotopic, and such an isotopy can be performed in the same Sobolev class H^ℓ. Consequently, $\mathcal{D}^\ell_0$ is the connected component of the identity in $\mathcal{D}^\ell$ (see [1], [16], [17]).

We first consider the case where S is the unit disc $D = \{z \in \mathbb{C} : |z| \leq 1\}$. In that case $\mathcal{D}^\ell = \mathcal{D}^\ell_0$. Furthermore, $\mathcal{D}^\ell_0$ can be retracted onto the space of those diffeomorphisms that fix $0 \in D$, and the latter space can be retracted onto the space of conformal automorphisms of D fixing 0. As that space is S^1, we see that S^1 is a deformation retract of $\mathcal{D}^\ell$, if S is the unit disc. Similarly, if S is the 2−sphere S^2, we get S^2 as a deformation retract of $\mathcal{D}^\ell$. If S is an annulus, we again get S^1 as deformation retract of $\mathcal{D}^\ell_0$, and for a torus T, we obtain T itself. In other words, in all those cases, $\mathcal{D}^\ell_0$ retracts onto the space of conformal automorphisms homotopic to the identity of S, and this continues to hold in the hyperbolic case. In the latter case, however, any conformal automorphism homotopic to the identity is the identity itself, and thus $\mathcal{D}^\ell_0$ retracts to a point, i.e. is contractible. If $\Gamma(S)$ denotes the group of homotopy classes, or, what is the same by Baer's theorem, of isotopy classes of diffeomorphisms, then the discrete group $\Gamma(S)$ is the group of connected components of $\mathcal{D}^\ell$, and in the hyperbolic case, $\mathcal{D}^\ell$ thus is homotopically equivalent to $\Gamma(S)$. $\Gamma(S)$ is also called the **mapping class group** or the modular group. We shall write $\Gamma_{p,k}$ in place of $\Gamma(S)$, if S has genus p and k boundary curves.

It is important to note that $\mathcal{D}^{k+1}$ operates by isometries on $\mathcal{M}^k$. Namely, for $g \in \mathcal{M}^k$, $\varphi \in \mathcal{D}^{k+1}$

$$\varphi : (S, \varphi^* g) \to (S, g)$$

is an isometry by definition of $\varphi^* g$.

Since isometries preserve curvature, $\mathcal{D}^{k+1}$ maps $\mathcal{M}^k_c$ into itself. If g is a fixed point for the action of $\mathcal{D}^{k+1}$ on $\mathcal{M}^k_c$, then for some $\varphi \in \mathcal{D}^{k+1}$,

$$\varphi^* g = g,$$

i.e.

$$\varphi : (S, g) \to (S, g)$$

is an isometry of the constant curvature metric g.

If S is the unit disc D or the sphere S^2, then $\varphi \in SO(2)$ or $SO(3)$ resp. If S is an annulus or torus T, the space of those isometries that are homotopic to the identity is identified with S^1 or T, resp. Finally, if we are in the hyperbolic case, there are only finitely many isometries, and in fact any isometry homotopic to the identity has to be the identity itself. Thus, the action of $\mathcal{D}_0^{k+1}$ on $\mathcal{M}_{-1}^k$ is free of fixed points.

Using the above identification between constant curvature metrics and complex structures, the preceding directly translates into statements about conformal automorphisms, i.e. automorphisms of Riemann surfaces. A Riemann surface of genus ≥ 2 possesses only finitely many conformal automorphisms, and none except the identity itself that is homotopic to the identity. For positive curvature, this is a little different, as for D and S^2, the groups of conformal automorphisms are $S\ell(2,\mathbb{R})$ and $S\ell(2,\mathbb{C})$, resp., which are larger than the groups of isometries for the metrics of constant curvature 1.

We return to the hyperbolic case, and the action

$$\mathcal{M}_{-1}^k \times \mathcal{D}_0^{k+1} \to \mathcal{M}_{-1}^k.$$

One may show that every $g \in \mathcal{M}_{-1}^k$ is orbit equivalent to some $h \in \mathcal{M}_{-1}$, i.e. of class C^∞; this means that given $g \in \mathcal{M}_{-1}^k$, we may find $\varphi \in \mathcal{D}_0^{k+1}$ with $\varphi^* g \in \mathcal{M}_{-1}$. Thus, the space of orbits is independent of k.

Definition 2.5.1. Let S be a compact oriented surface of genus p with k boundary curves of hyperbolic type, i.e. such that its Schottky double has genus $q = 2p + k - 1 \geq 2$. Then the Teichmüller space $T_{p,k}$ is the space of orbits of the above action. If $p = 0, k = 1$, we define $T_{0,1}$ as a point, while for $p = 0, k = 2$, we define $T_{0,2}$ as the space of orbits of

$$\bar{\mathcal{M}}_0^\ell \times \mathcal{D}_0^{\ell+1} \to \bar{\mathcal{M}}_0^\ell$$

The moduli space $M_{p,k}$ is obtained by replacing $\mathcal{D}_0$ by $\mathcal{D}$ in the above construction.

We also define the mapping class group or modular group $\Gamma_{p,k}$ as the group of homotopy classes of diffeomorphisms $\varphi : S \to S$, i.e. $\Gamma_{p,k} := \mathcal{D}^{\ell+1}/\mathcal{D}_0^{\ell+1}$ (again, this is independent of ℓ, as any homotopy class contains a C^∞ diffeomorphism).

From the definition, we have

$$M_{p,k} = T_{p,k}/\Gamma_{p,k}.$$

The moduli space is supposed to parametrize the different conformal structures on a surface S of type (p,k). Here, two conformal structures are considered different if there does not exist a conformal diffeomorphism between them. The advantage of working with Teichmüller space $T_{p,k}$ instead of moduli space $M_{p,k}$ stems from the fact that $T_{p,k}$ is obtained by dividing $\mathcal{M}_{-1}^k$ by an action without fixed points. For this reason, $T_{p,k}$ becomes a manifold, while $M_{p,k}$ has singularities. One has the following

Theorem 2.5.1. *For $q = 2p + k - 1 \geq 1$, $\mathcal{M}^\ell$ is homeomorphic to $T_{p,k} \times \mathcal{D}_0^{\ell+1} \times \mathcal{C}^\ell$, and $\mathcal{M}$ is diffeomorphic to $T_{p,k} \times \mathcal{D}_0 \times \mathcal{C}$.*

As $\mathcal{M}^\ell$ is convex, hence contractible, we conclude that $\mathcal{M}_{-1}^\ell, \mathcal{D}_0^{\ell+1}$, and $T_{p,k}$ are contractible as well.

2.6 Infinitesimal decompositions of metrics

We now wish to study the above constructions on the infinitesimal level. Given a metric $g = (g_{ij}) \in \mathcal{M}^k$, the conformal changes correspond to multiplications by positive functions. We thus decompose elements h of the tangent space $T_g\mathcal{M}^k$ as

$$h = \rho g + h', \tag{2.6.1}$$

where h' is trace free. As we have seen, this decomposition is orthogonal w.r.t. the natural Riemannian metric on $T_g\mathcal{M}^k$. We next consider the infinitesimal action of the diffeomorphism group. For that purpose, let $(\varphi_t) \subset \mathcal{D}^{k+1}, \varphi_0 = id$, be a smooth family of diffeomorphisms, generated by the vector field

$$V(z) := \frac{d}{dt}\varphi_t(z)_{|t=0}. \tag{2.6.2}$$

The infinitesimal change of the metric g under (φ_t) then is given by the Lie derivative

$$L_V g = \frac{d}{dt}(\varphi_t^* g)_{|t=0}. \tag{2.6.3}$$

With ∇ denoting the covariant derivative for the metric g,

$$\begin{aligned} \frac{d}{dt}((\varphi_t^* g)_{|t=0})_{ij} &= g_{ik}\left(\nabla_{\frac{\partial}{\partial z^j}} V\right)^k + g_{jk}\left(\nabla_{\frac{\partial}{\partial z^i}}\right)^k \\ &= g_{ij,k}V^k + g_{ik}V^k_{z^j} + g_{jk}V^k_{z^i}. \end{aligned} \tag{2.6.4}$$

In the above decomposition of $\mathcal{M}^\ell$, the directions corresponding to $\mathcal{C}^\ell$ are given by the tensors ρg, whereas those representing $\mathcal{D}_0^{\ell+1}$ are of the form $L_V g$. It remains to identify the Teichmüller directions, i.e. those corresponding to the factor $T_{p,k}$, as those that are orthogonal to the preceding two types.
Our computations simplify considerably if we use conformal coordinates so that the metric (g_{ij}) is of the form

$$g_{ij}(z) = \lambda^2(z)\delta_{ij}. \tag{2.6.5}$$

If a symmetric tensor h'' is orthogonal to all multiples ρg of g, it has to be trace free as already noted above. If it is orthogonal to all tensors $L_V g$, we get, using the symmetry of h''

$$\begin{aligned} 0 &= \int g^{ij} g^{kl} h''_{ik}(g_{j\ell,m}V^m + 2g_{jm}V^m_{z^\ell})\sqrt{\det g}\, dz^1\, dz^2 \\ &= \int \frac{1}{\lambda^2} h''_{ik}\left(\delta_{ik}\left(\frac{\partial}{\partial z^m}\lambda^2\right)V^m + 2\lambda^2 V^i_{z^k}\right) dz^1\, dz^2 \\ &= \int 2h''_{ik}V^i_{z^k}\, dz^1\, dz^2, \quad \text{since } h'' \text{ is traceless.} \end{aligned}$$

If this holds for all vector fields V, we conclude

$$\frac{\partial}{\partial z^k} h''_{ik} = 0 \quad \text{for } i = 1, 2. \tag{2.6.6}$$

This means that h''_{ik} is divergence free.
Of course, this condition is obtained through an integration by parts, and we also get a boundary term that needs to vanish. The requirement that the diffeomorphisms φ_t all map ∂S into itself implies that V has to be tangential to ∂S. If we choose our conformal coordinates $z = z^1 + iz^2$ near ∂S such that z^1 is tangential and z^2 normal to ∂S, we get the boundary term

$$\int_{\partial S} h''_{12} V^1 \, dz^1$$

and from its vanishing for all V

$$h''_{12} = 0 \quad \text{along} \quad \partial S. \tag{2.6.7}$$

Thus, h'' is symmetric, trace free, and divergence free, and its off diagonal component vanishes along the boundary. These conditions can be interpreted in a more concise manner as follows:
Being symmetric and trace free, h'' is of the form

$$\begin{pmatrix} h''_{11} & h''_{12} \\ h''_{12} & h''_{22} \end{pmatrix} \quad =: \quad \begin{pmatrix} u & v \\ v & -u \end{pmatrix}.$$

Being divergence free, this tensor then has to satisfy

$$u_{z^1} = -v_{z^2}, \; u_{z^2} = v_{z^1}.$$

Thus, $u - iv$ is holomorphic, or, as a tensor,

$$\begin{aligned} h'' &= u(dz^1)^2 - u(dz^2)^2 + 2v \, dz^1 \, dz^2 \\ &= \operatorname{Re}\left((u - iv)(dz^1 + dz^2)^2\right) \end{aligned} \tag{2.6.8}$$

is the real part of a holomorphic quadratic differential

$$\phi \, dz^2 = (u - iv) \, dz^2.$$

The boundary condition in this notation is

$$v = 0 \quad \text{along} \quad \partial S, \tag{2.6.9}$$

i.e.

$$\phi \, dz^2 \quad \text{is real on} \quad \partial S.$$

Thus, we have identified the tangent directions of $\mathcal{M}^\ell$ that correspond to $T_{p,k}$ as the real parts of holomorphic quadratic differentials on the Riemann surface defined by (S,g) that are real on the boundary ∂S .

It is also very instructive to evaluate our above Riemannian product on such holomorphic quadratic differentials

$$\phi_1\, dz^2 = (u_1 - iv_1)\, dz^2 \ , \ \phi_2\, dz^2 = (u_2 - iv_2)\, dz^2.$$

Using conformal coordinates as above, the result is

$$(\phi_1\, dz^2, \phi_2\, dz^2)_g = 2\int (u_1u_2 + v_1v_2)\frac{1}{\lambda^2(z)}\, dz\, d\bar{z} = 2 \text{ Re} \int \phi_1\bar{\phi}_2 \frac{1}{\lambda^2(z)}\, dz\, d\bar{z}. \tag{2.6.10}$$

This is nothing but the famous Weil-Petersson product on the space of holomorphic quadratic differentials on (S,g).
The preceding corresponds to a decomposition

$$h = L_V g + \sigma g + h'' \tag{2.6.11}$$

with h'' trace free and divergence free.
If we wish to compare this with (2.6.1), we need to split $L_V g$ into its trace part and its trace free part.
We recall from (2.6.3), (2.6.4)

$$(L_V g)_{ij} = g_{ik}\left(\nabla_{\frac{\partial}{\partial z^j}} V\right)^k + g_{jk}\left(\nabla_{\frac{\partial}{\partial z^i}} V\right)^k. \tag{2.6.12}$$

The trace part of $L_V g$ thus is

$$g^{ij}(L_V g)_{ij} = 2\left(\nabla_{\frac{\partial}{\partial z^m}} V\right)^m, \tag{2.6.13}$$

and the trace free part is as in (2.4.4)

$$\frac{1}{2}(\delta_i^k\delta_j^\ell + \delta_i^\ell\delta_j^k - g_{ij}g^{k\ell})(L_V g)_{k\ell} = (\delta_i^k\delta_j^\ell + \delta_i^\ell\delta_j^k - g_{ij}g^{k\ell})g_{km}\left(\nabla_{\frac{\partial}{\partial z^\ell}} V\right)^m =: P(V)_{ij}. \tag{2.6.14}$$

Comparing (2.6.1) and (2.6.11), we thus get

$$\rho = \sigma + 2\left(\nabla_{\frac{\partial}{\partial z^m}} V\right)^m \tag{2.6.15}$$

$$h' = h'' + P(V). \tag{2.6.16}$$

Our aim is to eliminate the degrees of freedom represented by σ and V from our functional integral. Thus, according to the Faddeev-Popov procedure, we compute the determinant

$$\left| \frac{\partial(h', \rho)}{\partial(V, \sigma)} \right| = \left| \det \begin{pmatrix} P & * \\ 0 & 1 \end{pmatrix} \right| = |\det P|, \tag{2.6.17}$$

in a notation that at this point is only symbolic.

The differential operator P maps vector fields into traceless symmetric tensors. From the derivation of (2.6.6), we recall the formula

$$\begin{aligned} (h', L_V g)_g &= (h', P(V))_g \quad \text{for symmetric, traceless} \quad h' \\ &= (-2\mathrm{div}_g h', V)_g, \end{aligned} \tag{2.6.18}$$

where in isothermal coordinates as in (2.6.5)

$$(\mathrm{div}_g h')_i = \frac{1}{\lambda^2} \frac{\partial}{\partial z^k} h'_{ik}, \tag{2.6.19}$$

and in general coordinates

$$(\mathrm{div}_g h')_i = \frac{1}{(\det g)^{\frac{1}{2}}} \frac{\partial}{\partial z^j} ((\det_g)^{\frac{1}{2}} g^{jk} h'_{ik}). \tag{2.6.20}$$

Thus, the adjoint operator of P is

$$P^* = -2\mathrm{div}, \tag{2.6.21}$$

but, of course, this needs to be supplemented by boundary conditions. To derive this boundary condition, we employ again isothermal coordinates as in (2.6.5). As before, near the boundary ∂S, we use coordinates $z = x + iy$ for which x is tangential and y normal to ∂S; i.e. ∂S is locally represented as $y = 0$. Writing

$$V = V^1 \frac{\partial}{\partial x} + V^2 \frac{\partial}{\partial y},$$

we get as the first boundary condition

$$V^2 = 0 \quad \text{for} \quad y = 0, \tag{2.6.22}$$

because all diffeomorphisms are required to map ∂S into itself, and so all vector fields have to be tangential to ∂S.

Next, for the equality

$$\int h'_{ij} P(V)_{ij} \frac{1}{\lambda^2} \, dz^1 \, dz^2 = -2 \int \left(\frac{\partial}{\partial z^j} h'_{ij} \right) V^i \, dz^1 \, dz^2 \tag{2.6.23}$$

(here, $z^1 = x, z^2 = y$), we need

$$h'_{12} = 0 \quad \text{for} \quad y = 0. \tag{2.6.24}$$

For $h' = P(V)$, i.e.

$$h'_{ij} = P(V)_{ij} = g_{ij,k}V^k + \lambda^2 V^i_{z^j}, \tag{2.6.25}$$

this becomes, using (2.6.22)

$$\frac{\partial}{\partial y} V^1 = 0 \quad \text{for} \quad y = 0. \tag{2.6.26}$$

Thus, the boundary conditions in terms of V are $V^2 = 0$ and $\frac{\partial}{\partial y} V^1 = 0$ for $y = 0$.
Conversely, for $V = -2 \operatorname{div} h'$, i.e.

$$V^i = -2 \frac{\partial}{\partial z^j} h'_{ij} =: -2h'_{ij,j}, \tag{2.6.27}$$

(2.6.22) becomes, using (2.6.24)

$$h'_{22,2} = 0 \quad \text{for} \quad y = 0, \tag{2.6.28}$$

or equivalently, as h' is trace free

$$h'_{11,2} = 0 \quad \text{for} \quad y = 0. \tag{2.6.29}$$

Thus, the boundary conditions in terms of h' are $h'_{12} = 0$ and $h'_{11,2} = 0$ for $y = 0$.
In both sets of boundary conditions, we have one Dirichlet and one Neumann condition, and these are exchanged by the operators P and P^*. These boundary conditions may be interpreted as symmetry conditions for making the transition from S to its Schottky double possible.

2.7 Complex analytic aspects

In order to discuss the preceding concepts and results in complex analytic terms, we need to recall some notions from the theory of Riemann surfaces [16].
In the beginning of the present §, we let Σ denote a compact Riemann surface without boundary. z will denote a local holomorphic coordinate.

Definition 2.7.1. A (holomorphic) line bundle L on Σ is given by an open covering $\{U_i\}_{i=1,\dots,m}$ of Σ and nonvanishing holomorphic transition functions $g_{ij} : U_i \cap U_j \to \mathbb{C}$ that satisfy

(i) $g_{ij}g_{ji} \equiv 1$ on $U_i \cap U_j$ for all i, j

(ii) $g_{ij}g_{jk}g_{ki} \equiv 1$ on $U_i \cap U_j \cap U_k$ for all i, j, k.

The geometric interpretation is that L is a collection of complex lines indexed by the points of Σ that vary holomorphically over Σ, that are locally trivial over each U_i (meaning that $L_{|U_i}$ is biholomorphic to $U_i \times \mathbb{C}$) and where for $z \in U_i \cap U_j$ one identifies the fiber $\{z\} \times \mathbb{C}$ in $U_i \times \mathbb{C}$ with the fiber $\{z\} \times \mathbb{C}$ in $U_j \times \mathbb{C}$ via multiplying all elements by the transition function $g_{ij}(z)$.

Definition 2.7.2. Let L be a line bundle with transition functions g_{ij}. A holomorphic section h of L is given by a collection of holomorphic functions $h_i = U_i \to \mathbb{C}$, $i = 1, ..., m$ that satisfy

$$h_i = g_{ij} h_j \quad \text{on} \quad U_i \cap U_j, \quad \text{for all} \quad i, j.$$

Thus, a holomorphic section is given by a holomorphic function for each local trivialization $U_i \times \mathbb{C}$ that transforms according to the rules defining the line bundle. Later on we shall also have occasion to investigate sections of a line bundle that are not necessarily holomorphic. In that case, we impose the above transformation behavior without the requirement that the h_i be holomorphic.

Definition 2.7.3. Let L be a line bundle with transition functions g_{ij}. A Hermitian metric λ^2 on L is given by a collection of smooth positive real valued functions λ_i^2 on U_i with

$$\lambda_j^2 = \lambda_i^2 \, g_{ij} \, \bar{g}_{ij} \quad \text{on} \quad U_i \cap U_j, \quad \text{for all} \quad i, j.$$

The norm of a section h of L then is locally defined by

$$\|h\|^2 = \lambda_i^2 h_i h_i \quad \text{on} \quad U_i.$$

The first Chern form of L w.r.t the metric λ^2 is defined as

$$c_1(L, \lambda^2) := \frac{1}{2\pi i} \frac{\partial^2}{\partial z \partial \bar{z}} \log \lambda_i^2 \, dz \wedge \, d\bar{z} \quad \text{on} \quad U_i.$$

(This is well defined since $\frac{\partial^2}{\partial z \partial \bar{z}} \log \, g_{ij} \bar{g_{ij}} = \frac{\partial}{\partial z} \left(\frac{\partial}{\partial \bar{z}} \log g_{ij} \right) + \frac{\partial}{\partial \bar{z}} \left(\frac{\partial}{\partial z} \log \bar{g_{ij}} \right) = 0$ as g_{ij} is a nonvanishing holomorphic function).

Remark. Nontrivial (i.e. not identically zero) holomorphic sections of a line bundle need not exist, because of the holomorphicity requirement. As there is no such holomorphicity requirement for metrics, those always exist by a simple partition of unity construction.

Since $c_1(L, \lambda^2)$ is exact on each U_i, it is closed:

$$dc_1(L, \lambda^2) = 0,$$

where d is the exterior derivative; here $dw = dz \wedge \frac{\partial}{\partial z} w + d\bar{z} \wedge \frac{\partial}{\partial \bar{z}} w =: \partial w + \bar{\partial} w$ (In fact, as $c_1(L, \lambda^2)$ is a 2−form on a surface, it is already closed for reasons of dimension, but the closedness of the first Chern form of a line bundle holds in any dimension.)
Therefore, the first Chern form defines a cohomology class, i.e. an element of $H^{1,1}(\Sigma, \mathbb{C})$.

Lemma 2.7.1. *The cohomology class defined by $c_1(L, \lambda^2)$ is independent of the choice of the Hermitian metric λ^2 on L.*

Proof. Let μ^2 be another metric, given by μ_i^2 on U_i. Then

$$\varphi := \frac{\lambda^2}{\mu^2} = \frac{\lambda_i^2}{\mu_i^2} \quad \text{on} \quad U_i$$

is a globally defined positive function on Σ. Then

$$\begin{aligned} c_1(L,\lambda^2) - c_1(L,\mu^2) &= \frac{1}{2\pi i}\frac{\partial^2}{\partial z \partial \bar{z}} \log \varphi \, dz \wedge d\bar{z} \\ &= \frac{1}{2\pi i} \partial\bar{\partial} \log \varphi \\ &= \frac{1}{2\pi i} d(\bar{\partial} \log \varphi) \qquad \text{(note } \partial^2 := \partial \circ \partial = 0 \\ &\qquad\qquad\qquad\qquad\qquad \text{and } \bar{\partial}^2 := \bar{\partial} \circ \bar{\partial} = 0) \end{aligned}$$

is exact, hence defines the trivial cohomology class. □

Definition 2.7.4. The cohomology class defined by $c_1(L,\lambda^2)$ (λ^2 a Hermitian metric on the line bundle L) is called the first Chern class of L and denoted by

$$c_1(L) \in H^{1,1}(\Sigma, \mathbb{C}).$$

The degree of the line bundle L is defined as

$$\deg L := \int_\Sigma c_1(L,\lambda^2) \qquad (= c_1(L)[\Sigma], \text{ if } [\Sigma] \in H_2(X,\mathbb{C})$$

denotes the so-called fundamental class of Σ).

Lemma 2.7.2. *A line bundle L of negative degree has no nontrivial holomorphic section. If the degree vanishes, nontrivial sections have no zeroes. If the degree is positive, it is given by the sums of the orders of the zeroes of any nontrivial holomorphic section.*

Proof. Let h be a holomorphic section of the line bundle L that is not identically zero. Then its zeroes are isolated, by holomorphicity; let they be denoted by $p_1, \ldots, p_k$, and let $B(p_j, r)$ be a small disc around p_j of radius r w.r.t. some local coordinate around p_j.
Then, for some Hermitian metric λ^2 on L

(2.7.1)
$$\frac{1}{2\pi i} \int\limits_{\Sigma \setminus \bigcup_{j=1}^k B(p_j,r)} \frac{\partial^2}{\partial z \partial \bar{z}} \log \lambda^2 dz \wedge d\bar{z} = \frac{1}{2\pi i} \int\limits_{\Sigma \setminus \bigcup B(p_j,r)} \frac{\partial^2}{\partial z \partial \bar{z}} \log \|h\|^2 dz \wedge d\bar{z},$$

as h is holomorphic and nonzero on $\Sigma \setminus \bigcup B(p_j, r)$, by the same argument as the one showing that $c_1(L,\lambda^2)$ is well defined.

Since $\|h\|^2$ is a function, we may integrate the last expression by parts to get (in local polar coordinates on each $B(p_j, r)$)

$$= \frac{1}{4\pi} \sum_{j=1}^{k} \int_{\partial B(p_j,r)} \frac{\partial}{\partial r} \log \|h\|^2 r d\vartheta.$$

Since, with $\|h\|^2 = \lambda_j^2 h_j \bar{h}_j$ locally,

$$\lim_{r \to 0} \int_{\partial B(p_j,r)} \frac{\partial}{\partial r} \log \lambda_j^2 \, r d\vartheta = 0 \quad \text{as} \quad \lambda_j^2 \quad \text{is positive}$$

and

$$\lim_{r \to 0} \int_{\partial B(p_j,r)} \frac{\partial}{\partial r} \log h_j \, r d\vartheta = 2\pi \mathrm{ord}_{p_j} h_j,$$

where the last expression denotes the order of the zero of h_j at p_j, we obtain, from letting $r \to 0$ in both sides of the above equation (2.7.1),

$$\deg L = \int_\Sigma c_1(L, \lambda^2) = \sum_j \mathrm{ord}_{p_j} h \geq 0.$$

□

Definition 2.7.5. For a line bundle L on Σ, we denote by $H^0(\Sigma, L)$ the complex vector space of its holomorphic sections, and $h^0(\Sigma, L) := \dim_{\mathbb{C}} H^0(\Sigma, L)$.

Corollary 2.7.1. *If* $\deg L < 0$, *then* $h^0(\Sigma, L) = 0$. □

Definition 2.7.6. Two line bundles L, L' with transition functions g_{ij} and g'_{ij} are isomorphic if there exists a nonvanishing holomorphic function φ_i on U_i for each i, such that

$$g'_{ij} = \frac{\varphi_i}{\varphi_j} g_{ij} \quad \text{on} \quad U_i \cap U_j.$$

In the sequel, we shall always identify isomorphic line bundles. For example, if h is a holomorphic section of L, then h', defined through

$$h'_i = \varphi_i h_i,$$

is a holomorphic section of L', and since φ_i is nonvanishing, the zeroes of h and h' coincide, and in fact this is all what counts about a holomorphic section. Of course, also the degree of isomorphic line bundles is the same.

If L, Λ are line bundles with transition functions g_{ij} and γ_{ij}, resp., we let $L^{-1} \otimes \Lambda$ be the line bundle with transition functions $g_{ij}^{-1} \gamma_{ij}$. In this way, we obtain the structure of an Abelian group on the set of (isomorphism classes of) line bundles, the trivial element being given by the trivial bundle $\Sigma \times \mathbb{C}$.

Definition 2.7.7. The Abelian group of line bundles on Σ is called the Picard group of $\Sigma, Pic(\Sigma)$.

Corollary 2.7.2. *Let L be a line bundle with a nowhere vanishing holomorphic section φ. Then L is isomorphic to the trivial line bundle $\Sigma \otimes \mathbb{C}$.*

Proof. If φ has no zeroes, then $g_{ij} = \frac{\varphi_i}{\varphi_j} \cdot 1$, where 1 of course can be interpreted as the transition function for the trivial bundle. Looking at Def.2.7.6 shows the assertion. □

Lemma 2.7.3. $\deg(L^{-1} \otimes \Lambda) = \deg \Lambda - \deg L$

Proof. If λ^2 is a metric on L, μ^2 one on Λ, then $\frac{\mu^2}{\lambda^2}$ is a metric on $L^{-1} \otimes \Lambda$, and

$$\begin{aligned} c_1\left(L^{-1} \otimes \Lambda, \frac{\mu^2}{\lambda^2}\right) &= \tfrac{1}{2\pi i}\partial\bar{\partial} \log \tfrac{\mu^2}{\lambda^2} \\ &= \tfrac{1}{2\pi i}\partial\bar{\partial} \log \mu^2 - \frac{1}{2\pi i}\partial\bar{\partial} \log \lambda^2 = c_1(\Lambda, \mu^2) - c_1(L, \lambda^2), \end{aligned}$$

and the assertion follows by integration. □

Definition 2.7.8. Let Σ be a Riemann surface covered by open sets U_i with local coordinates z_i. The canonical line bundle K of Σ then is defined by the transition functions

$$g_{ij} = \frac{\partial z_j}{\partial z_i} \quad \text{on} \quad U_i \cap U_j.$$

Thus, a holomorphic section h of the canonical bundle K has the transformation behavior

$$h_i = \frac{\partial z_j}{\partial z_i} h_j,$$

i.e. transforms as a holomorphic 1−form

$$\varphi(z)dz;$$

namely

$$\varphi_i(z_i)dz_i = \varphi_j(z_j)dz_j$$

precisely if

$$\varphi_i(z_i) = \varphi_j(z_j)\frac{\partial z_j}{\partial z_i}.$$

Thus, the holomorphic sections of K are the holomorphic 1−forms on Σ.
If λ^2 is a Hermitian metric on K, it transforms as

$$\lambda_i^2 = \lambda_j^2 \left(\frac{\partial z_j}{\partial z_i}\frac{\partial \bar{z}_j}{\partial \bar{z}_i}\right)^{-1}.$$

This is the inverse of the transformation behavior of a conformal metric on Σ, $\rho^2 dz d\bar{z}$.

Remark. So far, everything developed in this § extends from Riemann surfaces to complex manifolds of arbitrary dimension, with the exception of the notion of degree and the results depending on that notion, in particular Lemma 2.7.2. This point will be useful below when we shall have to deal with certain higher dimensional complex manifolds, namely (coverings of) moduli spaces of Riemann surfaces.

The central result in this context is the **Riemann-Roch theorem**.

Theorem 2.7.1. *Let Σ be a compact Riemann surface of genus p. Then for any line bundle L on Σ,*

$$h^0(\Sigma, L) - h^0(\Sigma, K \otimes L^{-1}) = \deg L - p + 1. \tag{2.7.2}$$

For the proof, we refer e.g. to [16], or to any other textbook on Riemann surfaces.
If L is the trivial line bundle $\Sigma \times \mathbb{C}$, then its holomorphic sections become holomorphic functions on Σ, and so they are precisely the constant functions. We have in that case

$$H^0(\Sigma, L) \cong \mathbb{C}, \quad \text{hence} \quad h^0(\Sigma, L) = 1.$$

The Riemann-Roch theorem then gives

Corollary 2.7.3. $h^0(\Sigma, K) = p$. □

Instead of the trivial bundle, we now insert the bundle $L = K$ into (2.7.2) to get in conjunction with the previous corollary

Corollary 2.7.4. $\deg K = 2p - 2$. □

Corollary 2.7.5. *For any line bundle L, $\deg L \in \mathbb{Z}$, and $\deg : Pic(\Sigma) \to \mathbb{Z}$ is a homomorphism of Abelian groups.*

Proof. $\deg L \in \mathbb{Z}$ as all the other terms in (2.7.2) are integers. The homomorphism property follows from Lemma 2.7.3. □

Corollary 2.7.6. *Let $Q(\Sigma)$ be the vector space of holomorphic quadratic differentials on the compact Riemann surface Σ of genus p. Then*

$$\dim_{\mathbb{C}} Q(\Sigma) = \begin{cases} 0 & \text{if } p = 0 \\ 1 & \text{if } p = 1 \\ 3p - 3 & \text{if } p \geq 2 \end{cases} \tag{2.7.3}$$

Proof. A holomorphic quadratic differential $\varphi(z)dz^2$ by its transformation behavior is a section of $K^2 (:= K \otimes K)$, the square of the canonical bundle. From Corollary 2.7.4 and Lemma 2.7.3,

$$\deg K^2 = 4p - 4. \tag{2.7.4}$$

If $p = 0$, this is negative, and so K^2 has no holomorphic sections by Lemma 2.7.2. If $p = 1$, $h^0(\Sigma, K) = 1$ by Corollary 2.7.3, and so K has a holomorphic section $\psi(z)dz$. Since $\deg K = 0$ in this case by Corollary 2.7.4, $\psi(z)dz$ has no zeroes by Lemma 2.7.2. Thus, $\psi(z)^2 dz^2$ defines a holomorphic quadratic differential, i.e. a section of K^2, without zeroes. By Corollary 2.7.2, K^2 is the trivial bundle, and its space of holomorphic sections is 1−dimensional.
Finally, let $p \geq 2$. Then

$$\begin{aligned} \deg(K^{-1}) &= 2 - 2p \quad \text{by Lemma 2.7.3 and Corollary 2.7.4} \\ &< 0, \end{aligned}$$

hence

$$h^0(K^{-1}) = 0 \quad \text{by Lemma 2.7.2.}$$

Riemann-Roch applied to $L = K^2$ then yields with (2.7.4)

$$h^0(K^2) = 3p - 3.$$

□

The inverse bundle of the canonical bundle K, K^{-1}, has sections transforming as

$$h(z)\frac{\partial}{\partial z},$$

i.e. as holomorphic vector fields. (Actually, K^{-1} can be identified with the holomorphic tangent bundle $T^{1,0}_{\mathbb{C}}\Sigma$, but we shall not need this fact in the sequel.) We now let $\bar{K}$ be the line bundle whose transition functions are the complex conjugate ones of those of K. Thus, $\bar{K}$ is not a holomorphic line bundle, but an antiholomorphic one. Its sections transform as

$$h(z)d\bar{z}.$$

We then have the $\bar{\partial}$-operators

$$\begin{aligned} \bar{\partial}_n : K^n &\to K^n \otimes \bar{K} \\ \varphi dz^n &\mapsto \varphi_{\bar{z}} dz^n \otimes d\bar{z} \end{aligned} \tag{2.7.5}$$

operating on - not necessarily holomorphic - sections of K^n. In fact, the holomorphic sections constitute precisely the kernel of $\bar{\partial}_n$:

$$\ker \bar{\partial}_n = H^0(\Sigma, K^n).$$

We now let

$$\rho^2 dz \otimes d\bar{z}$$

be a conformal metric on Σ. $\rho^2 dz \otimes d\bar{z}$ then is a nonvanishing section of the bundle

$$K \otimes \bar{K}.$$

If we drop the holomorphicity requirement from Def. 2.7.6, we get a notion of isomorphism of not necessarily holomorphic line bundles, and Cor. 2.5.2 and its proof continue to hold in that category. Thus, $K \otimes \bar{K}$ then is (isomorphic to) a trivial line bundle. As $K \otimes K^{-1}$ obviously is also trivial, we suspect that $\bar{K}$ and K^{-1} are isomorphic as line bundles; in fact, an isomorphism is given by

$$\begin{aligned} K^{-1} &\to \bar{K} \\ \frac{\partial}{\partial z} &\mapsto \rho^2 d\bar{z} \end{aligned} \tag{2.7.6}$$

on local sections.

With this isomorphism, $\bar{\partial}_n$ becomes an operator

$$\begin{aligned} \bar{\partial}_n : K^n &\longrightarrow K^{n-1} \\ \varphi(z) dz^n &\mapsto \frac{1}{\rho^2(z)} \varphi_{\bar{z}}(z) dz^{n-1} && \text{for} \quad n \geq 1 \\ \varphi(z) \left(\frac{\partial}{\partial z}\right)^{-n} &\mapsto \frac{1}{\rho^2(z)} \varphi_{\bar{z}}(z) \left(\frac{\partial}{\partial z}\right)^{-n+1} && \text{for} \quad n \leq 0. \end{aligned} \tag{2.7.7}$$

Defining

$$dz^n = \left(\frac{\partial}{\partial z}\right)^{-n}$$

we may achieve a uniform notation in the preceding formula.

On K^n, we have the Hermitian product

$$(\varphi dz^n, \psi dz^n) = \int_\Sigma \varphi(z) \bar{\psi}(z) (\rho^2(z))^{1-n} \frac{1}{2i} dz \wedge d\bar{z}. \tag{2.7.8}$$

W.r.t. this product, the adjoint of $\bar{\partial}_n$ then is

$$\begin{aligned} \bar{\partial}_n^* : K^{n-1} &\longrightarrow K^n \\ \psi dz^{n-1} &\mapsto -(\rho^2)^{n-1} \frac{\partial}{\partial z} \left((\rho^2)^{1-n} \psi\right) dz^n, \end{aligned} \tag{2.7.9}$$

since

$$\begin{aligned} \left(\frac{1}{\rho^2} \varphi_{\bar{z}} dz^{n-1}, \psi dz^{n-1}\right) &= \int \varphi_{\bar{z}} \bar{\psi} (\rho^2)^{1-n} \frac{1}{2i} dz \wedge d\bar{z} \\ &= -\int \varphi \frac{\partial}{\partial \bar{z}} ((\rho^2)^{1-n} \bar{\psi}) (\rho^2)^{n-1} (\rho^2)^{1-n} \frac{1}{2i} dz \wedge d\bar{z} \\ &= \left(\varphi dz^n, -(\rho^2)^{n-1} \frac{\partial}{\partial z} ((\rho^2)^{1-n} \psi) dz^n\right). \end{aligned}$$

The kernel of $\bar{\partial}_n^*$ thus consists of ψdz^{n-1} with

$$-(\rho^2)^{n-1} \frac{\partial}{\partial z} ((\rho^2)^{1-n} \psi) = 0 \tag{2.7.10}$$

Putting

$$V(z) := (\rho^2(z))^{1-n}\bar{\psi},$$

we then get a holomorphic $(n-1)$ vector field

$$V(z)\left(\frac{\partial}{\partial z}\right)^{n-1},$$

i.e. an element of

$$H^0(\Sigma, K^{1-n}).$$

Since the transition from ψdz^{n-1} to $V(z)\left(\frac{\partial}{\partial z}\right)^{n-1}$ involves the duality, thus

$$\operatorname{coker}\bar{\partial}_n := \ker\bar{\partial}_n^* = H^0(\Sigma, K^{1-n})^*, \tag{2.7.11}$$

and in particular,

$$\dim_{\mathbb{C}}\operatorname{coker}\bar{\partial}_n = h^0(\Sigma, K^{1-n}). \tag{2.7.12}$$

From the Riemann-Roch theorem, we thus obtain the index

$$\begin{aligned}\operatorname{ind}\bar{\partial}_n :&= \dim_{\mathbb{C}}\ker\bar{\partial}_n - \dim_{\mathbb{C}}\operatorname{coker}\bar{\partial}_n\\ &= h^0(\Sigma, K^n) - h^0(\Sigma, K^{1-n})\\ &= n\deg K - p + 1\\ &= (2n-1)(p-1) \quad \text{by Corollary 2.5.4}\end{aligned} \tag{2.7.13}$$

We now consider the case $p \geq 2$, leaving the considerations for $p = 0, 1$ to the reader.
For $p \geq 2$, by Corollary 2.5.4, Lemma 2.5.3, 2.5.1

$$h^0(\Sigma, K^{1-n}) = 0 \quad \text{for} \quad n > 1, \tag{2.7.14}$$

hence

$$\operatorname{ind}\bar{\partial}_n = \dim_{\mathbb{C}}\ker\bar{\partial}_n = (2n-1)(p-1) \quad \text{for} \quad p > 1, n > 1. \tag{2.7.15}$$

We are now ready to apply the preceding constructions and results and return to the differential operators encountered in previous §§.
First of all, we have

$$\begin{aligned}\bar{\partial}_0 : K^0 &\to K^{-1}\\ f(z) &\mapsto \frac{1}{\rho^2} f_{\bar{z}}(z)\frac{\partial}{\partial z}\end{aligned}$$

and

$$\begin{aligned}\bar{\partial}_0^* : K^{-1} &\to K^0\\ V(z)\frac{\partial}{\partial z} &\mapsto -\frac{1}{\rho^2}\frac{\partial}{\partial z}(\rho^2 V(z)),\end{aligned}$$

and we obtain the Laplace-Beltrami operator Δ from

$$\Delta f(z) = \frac{2}{\rho^2} f_{z\bar{z}}(z) = -2\bar{\partial}_0^* \bar{\partial}_0 f(z) \tag{2.7.16}$$

We have

$$\ker \bar{\partial}_0 = H^0(\Sigma, K^0) = \mathbb{C} \tag{2.7.17}$$

$$\operatorname{coker} \bar{\partial}_0 = \ker \bar{\partial}_0^* = H^0(\Sigma, K^1)^* = (\ker \bar{\partial}_1)^*, \tag{2.7.18}$$

i.e. the cokernel of $\bar{\partial}_0$ is the dual of the space of holomorphic $(1,0)$-forms.

We also wish to express the differential operator P from § 2.4 (see (2.6.14)) in the present notation. P was constructed from the operation of Lie derivative of the metric in the direction of a vector field, i.e. an infinitesimal diffeomorphism. Thus, we consider a diffeomorphism

$$z = \varphi(w)$$

and how it is pulling back the metric $\rho^2(z)dzd\bar{z}$:

$$\begin{aligned}\rho^2(\varphi(w))d\varphi(w)d\bar{\varphi}(w) \\ = \rho^2(\varphi(w))(\varphi_w\bar{\varphi}_w dw^2 + (\varphi_w\bar{\varphi}_{\bar{w}} + \varphi_{\bar{w}}\bar{\varphi}_w)dwd\bar{w} \\ + \varphi_{\bar{w}}\bar{\varphi}_{\bar{w}}d\bar{w}^2)\end{aligned}$$

If $z = \varphi_t(w)$ is a family of diffeomorphisms, depending smoothly on the parameter t, with $\varphi_0 = id, \dot{\varphi}_0 \left(= \frac{d}{dt}\varphi_{t|t=0}\right) = V$ as in § 2.4, we can compute

$$\begin{aligned}&\frac{d}{dt}(\rho^2(\varphi_t(w))d\varphi d\bar{\varphi})_{|t=0} \\ &= ((\rho^2 V)_z + (\rho^2 \bar{V})_{\bar{z}})dzd\bar{z} + \rho^2 \bar{V}_z dz^2 + \rho^2 V_{\bar{z}} d\bar{z}^2.\end{aligned}$$

The trace free part of this expression, i.e. the part defining P, is

$$P(V) = \rho^2 \bar{V}_z dz^2 + \rho^2 V_{\bar{z}} d\bar{z}^2 \tag{2.7.19}$$

$$= \bar{\partial}_2^*(\bar{V}dz) + \bar{\partial}_{-1}\left(V\frac{\partial}{\partial z}\right), \tag{2.7.20}$$

with the usual identification of K^{-2} and $\bar{K}^2$, under which $\frac{1}{\rho^2}V_{\bar{z}}\left(\frac{\partial}{\partial z}\right)^2$ becomes $\rho^2 V_{\bar{z}} d\bar{z}^2 \in K^2 \oplus \bar{K}^2$.
The adjoint then is written as

$$\begin{aligned}P^*(h) = P^*(h_1 dz^2 + h_2 d\bar{z}^2) = \bar{\partial}_2(h_1 dz^2) + \bar{\partial}_{-1}^*(h_2 d\bar{z}^2) \\ \in K^1 \oplus \bar{K}^1\end{aligned} \tag{2.7.21}$$

Since the two pieces occuring in (2.7.19) are complex conjugates of each other, it suffices to consider one of them, and so, with a change of notation,

$$P' : K^{-1} \to \bar{K}^2$$

(2.7.22) $$P'(V) = \bar{\partial}_{-1}\left(V\frac{\partial}{\partial z}\right)$$

$$P'^* : \bar{K}^2 \to K^1$$

(2.7.23) $$P'^*(h) = \bar{\partial}^*_{-1}(hd\bar{z}^2)$$

By (2.7.11), (2.7.14),

(2.7.24) $$\ker \bar{\partial}_{-1} = 0 \quad \text{for} \quad p \geq 2,$$

hence

(2.7.25) $$\ker P' = 0 \quad \text{for} \quad p \geq 2,$$

and likewise

(2.7.26) $$\operatorname{coker} P' = \operatorname{coker} \bar{\partial}_{-1} = H^0(\Sigma, K^2)^* = \ker(\bar{\partial}_2)^*,$$

meaning that the cokernel of P' is the dual of the space of holomorphic quadratic differentials.

We now return to the actual objects of interest for us, namely Riemann surfaces with boundary. We therefore change notation and, from now on, let Σ be a compact Riemann surface with boundary, and we denote its Schottky double by Σ'. Σ' thus is a closed Riemann surface to which the preceding constructions apply.

We then have an anticonformal involution

$$i : \Sigma' \to \Sigma'$$

whose fixed point set is $\partial\Sigma$, and with

$$\Sigma' = \Sigma \cup i(\Sigma),$$

i.e. Σ' consists of Σ and $i(\Sigma)$ glued together along their boundary.

$$H^0(\Sigma, K^n)$$

now is defined as the space of holomorphic sections of K^n, i.e. holomorphic n-differentials, that are real on $\partial\Sigma$.

Lemma 2.7.4.
$$H^0(\Sigma', K^n) = H^0(\Sigma, K^n) \otimes_{\mathbb{R}} \mathbb{C},$$

and in particular,

$$\dim_{\mathbb{R}} H^0(\Sigma, K^n) = \dim_{\mathbb{C}} H^0(\Sigma', K^n).$$

Proof. If $\phi \in H^0(\Sigma', K^n)$, then also

$$\bar{\phi} \circ i \in H^0(\Sigma', K^n)$$

Given $\phi \in H^0(\Sigma', K^n)$, we construct

$$\phi_1(z) := \phi(z) + \bar{\phi}(i(z))$$
$$\phi_2(z) := \phi(z) - \bar{\phi}(i(z))$$

ϕ_1 then is invariant under the reflection $\phi \mapsto \bar{\phi} \circ i$, and real on $\partial\Sigma$. Conversely, if ϕ_1 is a holomorphic differential that is real on $\partial\Sigma$, it can be reflected to a holomorphic differential on Σ'. From these observations, the assertions readily follow. □

Corollary 2.7.7. *$T_{p,k}$, the Teichmüller space of Riemann surfaces of genus p with k boundary curves, is a totally real submanifold of the Teichmüller space T_{2p+k-1} of closed Riemann surfaces of genus $2p+k-1$.*

Proof. $2p + k - 1$ is the genus of the Schottky double of a Riemann surface of genus p with k boundary curves. In § 2.3, we have identified the tangent space to $T_{p,k}$ at a Riemann surface Σ with the space of holomorphic quadratic differentials on Σ that are real on $\partial\Sigma$, i.e. with $H^0(\Sigma, K^2)$. Likewise, the tangent space to T_{2p+k-1} at Σ' is given by $H^0(\Sigma', K^2)$. The assertion then follows from the preceding lemma. □

We have already discussed the boundary conditions for the operator P above in § 2.4. For the operator Δ, let us recall that in § 2.2, we considered real valued functions vanishing on $\partial\Sigma$. Here, we look at complex valued functions $f = f^1 + if^2$ with

$$f^1 = 0 \quad \text{on} \quad \partial\Sigma,$$

and for reasons of holomorphicity, for f^2 we impose a Neumann condition, i.e.

$$f^2_y = 0 \quad \text{on} \quad \partial\Sigma,$$

if $\partial\Sigma$ is locally given by $\{y = 0\}$. Namely, if f is holomorphic with $f^1 = 0$, hence $f^1_x = 0$ on $\partial\Sigma$, then the Cauchy-Riemann equations imply $f^2_y = 0$. With these boundary conditions,

$$\bar{\partial}_0^* \bar{\partial}_0 : K^0 \to K^0$$

is self adjoint. Also

$$\ker \bar{\partial}_0 = i\mathbb{R}, \quad \text{the imaginary constants.}$$

As before, we write sections of $\bar{K}$ as

$$g(z)d\bar{z} = (g^1(z) + ig^2(z))d\bar{z}.$$

Then

$$\bar{\partial}_0^*(g d\bar{z}) = -\frac{1}{\rho^2} g_z dz d\bar{z}.$$

With the boundary conditions

$$g^1 = 0,\ g_y^2 = 0 \quad \text{on} \quad \partial\Sigma \quad (\text{again locally given by } \{y = 0\}),$$

$\bar{\partial}_0 \bar{\partial}_0^*$ is self adjoint as well.

The kernel of $\bar{\partial}_0^*$ thus consists of antiholomorphic 1-forms, i.e. complex conjugates of holomorphic 1-forms that are imaginary on $\partial\Sigma$, i.e. satisfy $g^1 = 0$ there. Again, the Neumann boundary condition $g_y^2 = 0$ then follows from the Cauchy-Riemann equations.

2.8 Teichmüller and moduli spaces of Riemann surfaces

The present § discusses some important global aspects of spaces of Riemann surfaces. While most of those results will not be directly utilized for our specific aim, namely the construction of the partition function for the Bosonic string, and thus can be omitted on a first reading, they should be quite helpful for putting the theory into a broader perspective. This S also makes contact with Teichmüller theory as originally developed by Ahlfors and Bers. Let S be a compact orientable differentiable surface of genus p. We have defined the Teichmüller space as the space of conformal classes of metrics on S divided by the action of the identity component of the diffeomorphism group. The moduli space is similarly obtained by dividing by the full diffeomorphism group. Traditionally, however, these spaces are defined as spaces of complex structures, and not as spaces of conformal classes of metrics. In abstract terms, this is the same, as each conformal class of metrics defines a complex structure on an orientable surface, and for each complex structure, one may construct a conformal metric. For genus $p \geq 2$, by Poincaré's uniformization theorem, one can view this also as a bijective correspondance between complex structures and hyperbolic metrics. In fact, this bijective correspondance yields a diffeomorphism between the spaces of (equivalence classes w.r.t the diffeomorphism group) conformal classes of metrics or hyperbolic metrics and (equivalence classes of) complex structures, and thus, the resulting Teichmüller spaces coincide as differentiable manifolds. This diffeomorphism, however, changes the type of tensors. In particular, while we have derived the space of holomorphic quadratic differentials on a given Riemann surface Σ as the tangent space to Teichmüller space at the point corresponding to Σ, for the space of equivalences classes of complex structures, it will become the cotangent space. The reason is that metrics and complex structures are given by different types of tensors. Namely, a metric is defined by a symmetric $(0, 2)$ tensor $(g_{ij}(z)\, dz^i \otimes dz^j)$ while an (almost) complex structure on a Riemann surface is given by a $(1, 1)$ tensor J. (On a Riemann surface, a complex structure is the same as an almost complex structure, because an integrability condition is automatically satisfied for dimensional reasons.) Thus, for each $z \in S$, $J(z)$ is a self-map of the tangent space $T_z S$,

$$J(z) : T_z S \to T_z S.$$

The condition for J to define an almost complex structure is

$$J^2(z) = -id_{|T_z S}, \qquad \text{for all } z, \tag{2.8.1}$$

which is simply an abstract version of the multiplication by i in complex analysis. Already at this level, we can see a direct relationship between metrics and almost complex structures. Namely, a metric $g_{ij}dz^i \otimes dz^j$ defines an almost complex structure J by

$$J_j^\ell = (\det g)^{\frac{1}{2}} g^{\ell k}(\delta_{1j}\delta_{2k} - \delta_{1k}\delta_{2j}). \tag{2.8.2}$$

In the other direction, (J_j^ℓ) determines a conformal class of metrics, i.e. a metric up to a conformal factor.

Now while a tangent direction to a space of metrics, i.e. of symmetric $(0, 2)$ tensors, is represented by a $(0, 2)$ tensor itself, a tangent direction to a space of almost complex structures, i.e. of $(1, 1)$ tensors with square being $-$identity, then has to be represented by a $(1, 1)$ tensor.

In fact, the two constructions are dual to each other in the sense that if we define Teichmüller space T_p as the space of equivalence classes of almost complex structures, then at a Riemann surface Σ, the space $Q(\Sigma)$ of holomorphic quadratic differentials on Σ represents the cotangent space of T_p at Σ, while the tangent space is given by the space $\mathcal{H}(\Sigma)$ of harmonic Beltrami differentials on Σ, i.e. objects of the form

$$\mu(z)\frac{\partial}{\partial z} \otimes d\bar{z} \quad \text{(Beltrami differential)} \tag{2.8.3}$$

satisfying, for a conformal metric $\rho^2(z)dz \otimes d\bar{z}$ on Σ,

$$-\frac{1}{\rho^2(z)}\frac{\partial}{\partial z}(\rho^2(z)\mu(z)) = 0 \quad \text{(harmonic)}. \tag{2.8.4}$$

Thus, holomorphic quadratic differentials $\psi(z)dz^2$ and harmonic Beltrami differentials $\mu(z)\frac{\partial}{\partial z} \otimes d\bar{z}$ correspond to each other via

$$\mu(z) = \frac{\bar{\psi}(z)}{\rho^2(z)}.$$

It may be a little surprising that we need to specify a metric on Σ in order to define a tangent direction to the space of almost complex structures, i.e. a space that does not involve individual metrics, but only conformal classes of metrics as explained above, while a tangent direction to the space of hyperbolic metrics, namely a holomorphic quadratic differential, did not involve a metric. We can explain this as follows. We consider the space $B(\Sigma)$ of all smooth Beltrami differentials $\mu(z)\frac{\partial}{\partial z}\otimes d\bar{z}$, not necessarily harmonic. A Beltrami differential deforms the conformal class of metrics respresented by

$$dz \otimes d\bar{z} \text{ (i.e. } g_{ij}dz^i \otimes dz^j = dz^1 \otimes dz^1 + dz^2 \otimes dz^2)$$

to

$$(dz + t\mu d\bar{z}) \otimes \overline{(dz + t\mu d\bar{z})}$$

for $t \in \mathbb{R}$ with $|t|$ small enough, and thus also the almost complex structure given by (2.8.2). (The reader is invited to derive a more direct formula for the relationship between J and μ by differentiating the relation (2.8.1) with respect to a variation j of J:

$$\frac{d}{dt}(J + tj)^2_{|t=0} = 0.)$$

As before, one needs to divide out the space $N(\Sigma)$ of those Beltrami differential that arise from infinitesimal diffeomorphisms of Σ, i.e. from vector fields $V(z)$ on Σ.

Now reversing the argument that characterized the holomorphic quadratic differentials as the directions orthogonal to the action of the diffeomorphism group, we see that a Beltrami differential $\mu\frac{\partial}{\partial z} \otimes d\bar{z}$ is of the form $V_{\bar{z}}\frac{\partial}{\partial z} \otimes d\bar{z}$, where $V\frac{\partial}{\partial z}$ is a vector field on Σ, precisely if

$$\int \phi dz^2 \wedge \mu\frac{\partial}{\partial z} \otimes d\bar{z} = \int \phi\mu dz \wedge d\bar{z} = 0 \quad \text{for all } \phi dz^2 \in Q(\Sigma),$$

i.e. if the Beltrami differential is orthogonal to all holomorphic quadratic differentials w.r.t the natural pairing between the two spaces.

Therefore $B(\Sigma)/N(\Sigma)$ represents the tangent space of the space of almost complex structures at Σ. If we are given a conformal metric $\rho^2 dz \otimes d\bar{z}$ on Σ, we may identify $B(\Sigma)/N(\Sigma)$ with the space $\mathcal{H}(\Sigma)$ of harmonic Beltrami differentials,

$$\mathcal{H}(\Sigma) = \{\frac{\bar{\phi}}{\rho^2}\frac{\partial}{\partial z} \otimes d\bar{z}, \phi\, dz^2 \in Q(\Sigma)\},$$

and this is the point where the choice of a metric comes into play.

In order to see that $\mathcal{H}(\Sigma)$ is the orthogonal complement of the space of ($\frac{\partial}{\partial \bar{z}}$-derivatives of) vector fields on Σ, we assume that $V\frac{\partial}{\partial z}$ is a vector field for which $V_{\bar{z}}\frac{\partial}{\partial z} \otimes d\bar{z}$ is a harmonic Beltrami differential. We then have

$$\int_\Sigma V_{\bar{z}}\, \bar{V}_z\, \rho^2 dz \wedge d\bar{z} = \int \frac{\partial}{\partial z}(\rho^2 V_{\bar{z}})\bar{V} dz \wedge d\bar{z} = 0,$$

where the integration by parts is allowed because $\bar{V}$ is a vector field (and so, $\bar{V}_z$ is a derivative). Thus,

$$V_{\bar{z}} = 0,$$

i.e. V is a holomorphic vector field. Since we assume that the genus p of Σ is at least 2, V has to vanish, see (2.7.24).
Having clarified the preceeding issue, we may wish to ask how the geometry of Teichmüller space depends on the choices of metrics $\rho^2 dz \otimes d\bar{z}$ on the Riemann surfaces Σ. In §§ 2.5, 2.6, we have employed the hyperbolic metric, but all the abstract constructions encountered so far are also meaningful for other choices of metrics. In particular, we can define the Weil-Petersson metric on $Q(\Sigma)$

$$(\phi_1 dz^2, \phi_2 dz^2) = 2 \text{ Re} \int \phi_1 \bar{\phi}_2 \, \frac{1}{\rho^2(z)} dz \wedge d\bar{z}$$

and by duality on $\mathcal{H}(\Sigma)$ as well

$$(\mu_1 \frac{\partial}{\partial z} \otimes d\bar{z}, \mu_2 \frac{\partial}{\partial z} \otimes d\bar{z}) = 2 \text{ Re} \int \mu_1 \bar{\mu}_2 \rho^2(z) dz \wedge d\bar{z}.$$

This defines a Riemannian metric on Teichmüller space, and one may ask about the properties of the resulting Riemannian manifold, for example about the completeness, the curvature properties of the metric or whether it happens to be a Kähler manifold w.r.t. some natural complex structure. All these properties should also pass down to the moduli space M_p (ignoring the technical issues of the singularities of that space for the sake of discussion).
For the issue of completeness, we need to discuss the possible degenerations of Riemann surfaces of genus p. It turns out that in the present context, only the following type of degeneration is relevant: Let Σ_0 be a Riemann surface of genus $p-1$, or the disjoint union of two such surfaces of genus p_1 and p_2, resp., with $p_1 + p_2 = p$. Take points x_1, x_2, the so-called punctures, on Σ_0, one in each component in the disconnected case, with disjoint neighborhoods U_1, U_2, admitting complex coordinates

$$z : U_1 \to D, \; w : U_2 \to D,$$

with $D = \{z \in \mathbb{C} : |z| < 1\}$ the unit disk, as usual. For $t \in D$, we then form a Riemann surface Σ_t by removing $\{|z| \leq |t|\}, \{|w| \leq |t|\}$ from U_1, U_2, resp., and identify z with w by the relation

$$zw = t$$

on the remaining parts. Σ_t then is a Riemann surface of genus p, and as $|t| \to 0$, it degenerates to Σ_0. The geometry of this degeneration can be nicely visualized if we equip all the surfaces Σ_t with their hyperbolic metrics. As $|t| \to \infty$, we obtain a noncompact, complete hyperbolic surface with two cusps corresponding to x_1, x_2, i.e. we get the standard hyperbolic metric on the punctured disk $D \backslash \{0\}$, given by

$$\frac{4}{|z|^2 (\log |z|^2)^2} \, dz \, d\bar{z},$$

via our coordinates on the punctured disks $U_1 \backslash \{x_1\}$ and $U_2 \backslash \{x_2\}$ in Σ_0.

From the point of view of hyperbolic geometry, one sees that this is the only possible type of degeneration by first using an argument of Mumford that the space of hyperbolic Riemann surfaces with a fixed positive lower bound for the lengths of closed geodesics represents a compact subset of moduli space $\mathcal{M}_p$ (but not of Teichmüller space T_p, because of the action of the mapping class group Γ_p). Having established that fact one then invokes the so-called collar lemma stating that any closed hyperbolic geodesic possesses a region about it isometric to the region on Σ_t obtained by gluing the annuli from U_1 and U_2 together. A detailed presentation is given in Jost [17] and [20].

We interpret this fact as saying that the moduli space M_p can be completed to become a compact space by adding as boundary elements Riemann surfaces of lower topological type, i.e. either of genus $p-1$, or disconnected with pieces whose genera p_1, p_2 satisfy $p_1 + p_2 = p$, together with distinguished points, punctures, nodes or whatever they are called, x_1, x_2.

Of course, in order to obtain a compactification of the moduli space, the procedure needs to be iterated until no further degeneration is possible. In that manner, one obtains a moduli space $\bar{M}_p$ that has also been constructed from a rather different perspective in algebraic geometry as the Mumford-Deligne moduli space of stable curves of genus p.

The question emerges whether these ideal boundary points that we have just constructed are at a finite or infinite distance from the interior w.r.t. the (Weil-Petersson)-metric on M_p. The answer will actually depend on the choice of metrics on the surfaces Σ. We treat the case of hyperbolic metrics here as the one having received by far the most attention in the literature. Namely, in particular, here we have the very precise asymptotic estimates of Masur which we now describe:

The tangent vector to the family $\{\Sigma_t\}_{t\in D}$ is represented by the Beltrami differentials ν_t given by

$$\frac{1}{2t\log|t|}\,\frac{z}{\bar{z}}\,\frac{d\bar{z}}{dz} = \frac{1}{2t\log|t|}\,\frac{w}{\bar{w}}\,\frac{d\bar{w}}{dw} \quad \text{on the collar} \quad A_t = \{|t| < |z| < 1\} \subset \Sigma_t$$

and

$$0 \text{ on the rest of } \Sigma_t.$$

These Beltrami differentials, however, are not yet harmonic, and according to our above discussion, we now wish to represent these directions by harmonic ones. For that purpose, we consider the following basis $\{\phi_{1,t,}, \ldots, \phi_{3p-3,t}\}$ of

$Q(\Sigma)$:

$$\begin{aligned}
\phi_{1,t}(z) &= -\frac{t}{\pi}\left(\frac{1}{z^2} + \sum_{k=-1}^{\infty} a_{1k}(t)z^k + \sum_{k=-\infty}^{-3} a_{1k}(t)t^{-k-2}z^k\right) \\
&= -t\left(\frac{1}{\pi z^2} + \alpha_{1,t}(z) + b_{1,t}\frac{t}{z^3} + \beta_{1,t}(\frac{t}{z})\frac{t^2}{z^4}\right) \\
\phi_{j,t}(z) &= \sum_{k=-1}^{\infty} a_{jk}(t)z^k + \sum_{k=-\infty}^{-3} a_{jk}(t)t^{-k-2}z^k \qquad \text{for } j > 1
\end{aligned}$$

The metric on the cotangent space of T_p, i.e. on $Q(\Sigma_t)$ w.r.t. this basis is denoted by

$$g_t^{i\bar{j}}.$$

Using the chosen metric ρ_t^2 on Σ_t, we then define harmonic Beltrami differentials by

$$\mu_{1,t} = \frac{\bar{\phi}_{1,t}}{\rho_t^2}$$

$$\mu_{j,t} = \frac{\bar{\psi}_{j,t}}{\rho_t^2}$$

where the $\psi_{j,t}$ are suitable linear combinations of the $\phi_{j,t}$, for $j \geq 2$, such that

$$\begin{aligned}
(\nu_t, \mu_{1,t}) &= 1 \\
(\nu_t, \mu_{j,t}) &= 0 \qquad \text{for } j \geq 2.
\end{aligned}$$

In order to check whether a boundary point is at finite distance from the interior, one has to verify whether

$$\int_0^{\varepsilon} g_{1\bar{1},t}^{\frac{1}{2}}\, d\tau = \int_0^{\varepsilon} ||\nu_t|| d\tau < \infty \qquad \text{for } \tau = |t|, \qquad \text{for some } \varepsilon > 0.$$

If the metrics ρ_t^2 on Σ_t are the hyperbolic ones, then this is finite indeed, and so the moduli space M_p is metrically incomplete. In fact, the above formulae contain more information: In order to describe that, we first observe that each boundary stratum, i.e. a space of either twice punctured surfaces of genus $p-1$ or of surfaces with two once punctured pieces of fixed genera p_1, p_2, satisfying $p_1 + p_2 = p$, is itself a moduli space with its own intrinsic Weil-Petersson metric. The positions of the punctures just yield two additional complex dimensions, so that each boundary stratum has complex dimension $3(p-1) - 3 + 2 = 3p - 4$. Thus, the boundary of M_p has complex codimension 1. Cotangent vectors of a boundary stratum corresponding to directions where the position of a puncture is moved then are given by meromorphic quadratic differentials with simple poles at the punctures. Returning to the above basis of $Q(\Sigma_t)$, we see that the term with coefficient $\frac{1}{z^2}$ has been singled out. On Σ_0, in the limit, this leads to a meromorphic quadratic differential with a double pole, corresponding to opening up the punctures in the direction of the family Σ_t transversal to the boundary stratum. The other holomorphic quadratic differentials converge to meromorphic quadratic differentials on Σ_0 with at most simple poles, and in fact the corresponding Weil-Petersson products converge to the corresponding products on the boundary stratum. Thus, everything fits together metrically.

As already indicated, also other canonical choices of metrics on the surfaces Σ in M_p are possible. One possibility is the Bergman metric

$$\rho_B^2\, dz d\bar{z} = \frac{1}{2p} \sum_{i=1}^{p} \vartheta i \wedge \bar{\vartheta}_i$$

where $(\vartheta_i)_{i=1,\dots,p}$ is a basis of the space of holomorphic 1-forms on Σ, chosen as L^2-orthogonal, i.e. satisfying

$$(\vartheta_i, \vartheta_j) := \frac{i}{2} \int \vartheta_i \wedge \bar{\vartheta}_j = \delta_{ij}.$$

For those knowing some algebraic geometry, this can also be described as the pull-back of the metric on the Jacobian J_Σ via the Albanese period map

$$\pi_\Sigma : \Sigma \to J_\Sigma.$$

The Jacobian J_Σ is a principally polarized Abelian variety, and the map that associates to each Σ its Jacobian J_Σ defines a map

$$\pi : T_p \to H_p$$

from Teichmüller space into the Siegel upper half space H_p, the moduli space of principally polarized Abelian varieties of dimension p, a Hermitian symmetric space of noncompact type. The map π is not surjective, because not every prinipally polarized Abelian variety is a Jacobian of some Riemann surface (except for $p = 1, 2$). It is injective, but not everywhere an immersion. Because of that, if one pulls back the symmetric metric on H_p via π to T_p, one only gets a metric with singularities on T_p. That singular metric is dominated by the Bergman metric. Also, there is a natural discrete group Λ_p acting on H_p that identifies isomorphic Abelian varieties with different polarizations, and this is compatible with the action of the mapping class group Γ_p on T_p (a principal polarization of the Jacobian J_Σ corresponds to the choice of a basis of holomorphic 1-forms on the Riemann surface Σ, and the action of the mapping class group permutes the basis vectors and thus changes the polarization). Therefore, we also get an induced map

$$\pi : M_p \to H_p/\Lambda_p =: N_p$$

Another canonical metric is the Arakelov metric $\gamma^2 dz d\bar{z}$ given by the formula

$$\frac{\partial^2}{\partial z \partial \bar{z}} \log \gamma = \pi(2 - 2p)\rho_B^2$$

the Arakelov metric can be expressed in terms of the Green function $G(z,w)$, defined by

$$\frac{\partial^2}{\partial z \partial \bar{z}} G(z,w) = \delta_w(z) - \rho_B^2$$

(δ_w being the delta distribution supported at w), normalized by

$$\int_\Sigma G(z,w)\rho_B^2 \frac{i}{2} dz \wedge d\bar{z} = 0.$$

We then have the relation

$$\log \gamma(z) = \lim_{w \to z} (G(z,w) - \log|z-w|).$$

For the Bergman metric, it turns out that a boundary stratum is at finite distance if Σ_0 is disconnected, but at infinite distance if it is connected, see Habermann-Jost [12]. The same holds for the metric on T_p pulled back via π from H_p. Here, this is easy to understand. Namely, if Σ_0 is disconnected, its genus $p_1 + p_2$ is still p, and so it still has p linearly independent holomorphic 1-forms, and its Jacobian still has dimension p and therefore lies in the interior of H_p. If however Σ_0 is connected, its genus is $p-1$, and so its Jacobian has lower dimension. The Abelian varieties of dimension less than p also correspond to the points in a compactification of the moduli space N_p of Abelian varieties, the Satake-Baily-Borel compactification. Since H_p and consequently also its quotient N_p are complete metric spaces, these boundary points are at infinite distance from the interior. While the boundary strata of the Mumford-Deligne compactification $\bar{M}_p$ of M_p have complex codimension 1, and the singularities of this space are rather mild (essentially quotient singularities corresponding to fixed points of Γ_p, i.e. Riemann surfaces with nontrivial automorphisms), and the various boundary strata intersect transversally, and as explained above, even the Weil-Petersson metric extends nicely, the situation for $\bar{N}_p$ is different. Boundary strata are of codimension 3 (except for $p = 2$, where the codimension is 2), and the singularities at the boundary are very complicated. Other, less singular compactifications have been constructed, but the Satake-Baily-Borel compactification in many regards is the most natural one. Extending π to the boundary of M_p induces a different compactification $\bar{\bar{M}}_p$ of M_p, studied by Baily [3]. Essentially, it corresponds to forgetting the positions of the punctures of the surfaces in the boundary of M_p. Since each puncture is determined by one complex parameter, this increases the codimension of the boundary by 2 (thus most boundary components have codimension 3, except for the one with $p_1 = p-1, p_2 = 1$, i.e. where we have a component containing once punctured tori, for which the codimension is only 2), and this collapse again makes the boundary highly singular. Nevertheless, as $\bar{M}_p$, the Baily compactification $\bar{\bar{M}}_p$ can also be constructed as an algebraic variety, essentially because we get an injective extension

$$\pi : \bar{\bar{M}}_p \to \bar{N}_p$$

into the algebraic variety $\bar{N}_p$.

Now for string theory, it is natural to ask for a moduli space of Riemann surfaces of all (finite) topological types, some kind of universal moduli space that is stratified by the moduli spaces for each fixed genus p. Such a space would constitute a natural space on which to perform the functional integration for the partition function of Bosonic string theory. Let us briefly discuss whether the results described so far allow such a construction. The starting point is the fact that the moduli space for surfaces of genus p is compactified by adding moduli spaces of lower topological type. Thus, in its boundary it contains a moduli space for surfaces of genus $p-1$ with two punctures which in turn in its boundary contains a moduli space of surfaces of genus $p-2$ with 4 punctures, and so on. Thus, inductively, for any $n \leq p$, we have a stratum of surfaces of genus $p-n$ with $2n$ punctures. Other strata contain disconnected surfaces, but again a stratum of codimension n corresponds to surfaces with $2n$ punctures. Thus, if we now let p get arbitrarily large, the number of punctures that can occur in some stratum also goes to infinity.

This problem does not occur for the Baily compactification $\bar{\bar{M}}_p$ of M_p. It simply contains the moduli space M_{p-1} and other strata in its boundary. Thus, it seems natural to construct a universal moduli space as the inductive limit of the spaces $\bar{\bar{M}}_p$ as p goes to infinity.

From the point of view of string scattering theory, however, punctures are still relevant. In the same way that in our approach, we consider Riemann surfaces with boundary, the boundary curves corresponding to initial and final states of a string, punctures of Riemann surfaces can be positioned at infinity by introducing complete hyperbolic metrics on their complements and then correspond to asymptotic states. This can be formalized by inserting vertex operators in our functional integrals, in a similar, but simpler manner as our treatment of the boundary terms below.

2.9 Determinants

The theory of determinant line bundles of elliptic operators was introduced by Quillen in [28].

Let $\mathcal{F} = \mathcal{F}(H_1, H_2)$ be the space of Fredholm operators $T : H_1 \to H_2$ between the complex Hilbert spaces H_1, H_2. Here, by definition, T has the Fredholm property if its kernel is finite dimensional and its range is closed and has a finite dimensional complement, the so-called cokernel. One defines the index of T as

$$\text{ind}\, T := \dim \ker T - \dim \operatorname{coker} T \tag{2.9.1}$$

$\mathcal{F}$ is an open subset of the Banach space of bounded linear operators, hence a complex Banach manifold, and the index is constant on each component.

For finite dimensional vector space V, e.g. $V = \ker T$ or $\operatorname{coker} T$, we put

$$\text{Det } V := \wedge^{\dim V}(V), \quad \text{the highest nontrivial exterior product.}$$

Then the determinant line of T is

$$\text{Det } T := (\text{Det } \ker T)^* \otimes \text{Det } \operatorname{coker} T \tag{2.9.2}$$

For a finite dimensional subspace W of H_2, let U_W be the open set of those $T \in \mathcal{F}$ for which

$$T(H_1) \oplus W = H_2 \tag{2.9.3}$$

(if this holds, W is called transversal to T).
The exact sequence

$$0 \to \ker T \to T^{-1}W \xrightarrow{T} W \to \operatorname{coker} T \to 0 \tag{2.9.4}$$

yields a canonical isomorphism

$$\text{Det } T = (\text{Det } \ker T)^x \otimes \text{Det } \operatorname{coker} T \cong (\text{Det } T^{-1}W)^* \otimes \text{Det } W \tag{2.9.5}$$

by identifying for each $v \in T^{-1}W$ with $Tv \neq 0$

$$1 \cong v^* \otimes Tv. \tag{2.9.6}$$

The family of subspaces $T^{-1}W$ forms a holomorphic vector bundle over the open set $U_W \subset \mathcal{F}$, and so the rhs of (2.9.5) forms a holomorphic line bundle over U_W. Requiring (2.9.5) to be an isomorphism of holomorphic line bundles over U_W for any W, Det becomes a holomorphic line bundle over $\mathcal{F}$, the so-called determinant line bundle.
T induces a map

$$\det T : \text{Det } T^{-1}W \to \text{Det } W,$$

i.e. an element

$$\det T \in (\text{Det } T^{-1}W)^* \otimes \text{Det } W \tag{2.9.7}$$

which is nonzero precisely if T is invertible. Under the isomorphism (2.9.5), (2.9.6), in the invertible case, Det T is identified with $\mathbb{C}$, and $\det T$ with $1 \in \mathbb{C}$. $\det T$ defines a continuous, in fact a holomorphic section of the determinant line bundle vanishing precisely at those T that are not invertible.

We now consider the more special case of elliptic differential operators that will allow us to define a metric on the determinant line bundle. We let Y be a manifold - in fact, it will be a Kähler manifold later on - parametrizing a smooth - later holomorphic - family $\{D_y\}_{y \in Y}$ of elliptic differential operators

$$D_y : \Gamma(E_y^1) \to \Gamma(E_y^2)$$

where E_y^1, E_y^2 are Hermitian vector bundles (over some Kähler manifold M_y), and $\Gamma(E)$ denotes the space of (smooth) sections of a bundle E. The Hermitian metrics on E_y^1, E_y^2 induce L^2-products on the spaces of sections by integration, and with the help of these Hermitian products, we form the adjoints D_y^* and the Laplacians

$$\Delta_y := D_y^* D_y : \Gamma(E_y^1) \to \Gamma(E_y^1).$$

Ker D_y and Ker D_y^* then also carry induced Hermitian L^2-products. We let ζ_y be the zeta function of the elliptic operator Δ_y; for Re $s > \dim_{\mathbb{C}} M_y$, we have

$$\zeta_y(s) = \sum \lambda_n^{-s}, \tag{2.9.8}$$

where λ_n runs over the nonvanishing eigenvalues of Δ_y (always counted with multiplicity). From the Weyl estimates for the asymptotic growth of the eigenvalues of elliptic operators, one knows (at least in those cases that will be considered below) that ζ_y admits a meromorphic extension to all of $\mathbb{C}$ that is holomorphic at 0. The reason for this are asymptotic estimates for the growth of eigenvalues of the type first proved by H. Weyl for the Dirichlet problem for the Laplacian in domains in Euclidean space $\mathbb{R}^d$; namely the eigenvalues λ_n grow proportionally to $(\frac{n}{Vol(M)})^{\frac{2}{d}}$ up to terms of lower order where $Vol(M)$ is the volume of the domain or manifold under consideration. $\exp(-\zeta_y'(0))$ is interpreted as the determinant of Δ_y acting on the orthogonal complement of $\ker D_y$. (The formal reason for this is that if we had only finitely many λ_n then we would have the identity $\frac{d}{ds}(\Sigma\lambda_n^{-s})_{|s=0} = -\Sigma \log \lambda_n$, hence $\exp\left(\frac{d}{ds}(\Sigma\lambda_n^{-s})_{|s=0}\right) = \Pi\lambda_n$.) We then define an inner product on Det $D =$ (Det $\ker D)^* \otimes$Det $\ker D^*$ by multiplication of the one induced by the products on $\ker D_y$ and $\ker D_y^*$ by $\exp(-\zeta_y'(0))$. If we choose orthonormal bases for these kernels and take the exterior product of all basis elements, we obtain a nonzero $v \in$ Det D that is determined up to a factor of absolute value 1 and we have

$$\|v\|_Q^2 = \exp(-\zeta_y'(0)). \tag{2.9.9}$$

Lemma 2.9.1. *These inner products on the family of lines Det D_y define a smooth inner product on the determinant line bundle on $\mathfrak{F}$, the so-called Quillen product.*

Proof. We drop the subscript y for the moment for notational simplicity. We let E_λ^1, E_λ^2 be the spaces spanned by eigensections of D^*D and DD^*, resp., with eigenvalues $< \lambda$. From our above considerations, we get a canonical isomorphism

$$\text{Det } D \cong (\text{Det } E_\lambda^1)^* \otimes \text{Det } E_\lambda^2,$$

because if $D^*D\varphi_n = \lambda_n\varphi_n$, with $\lambda_n > 0$, i.e. φ_n is an eigensection of D^*D with nonvanishing eigenvalue, then $\psi_n := D\varphi_n$ satisfies $DD^*\psi_n = \lambda_n\psi_n$, i.e. ψ_n is an eigensection of DD^* with the same eigenvalue, and so E_λ^1 and E_λ^2 differ precisely by the kernels of D and D^*.

If $\varphi_1, ..., \varphi_{k(\lambda)}$ and $\psi_1, ..., \psi_{\ell(\lambda)}$ are bases of E^1_λ and E^2_λ, resp., we obtain a section s_λ of

$$L_\lambda := (\text{Det } E^1_\lambda)^* \otimes \text{Det } E^2_\lambda$$

as

$$s_\lambda(\varphi_1 \wedge ... \wedge \varphi_{k(\lambda)}) = \sigma_\lambda \psi_1 \wedge ... \wedge \psi_{\ell(\lambda)}$$

with induced norm

$$\|s_\lambda\|^2_{\text{ind}} = |\sigma_\lambda|^2 \frac{\det(< \psi_\alpha, \psi_\beta >)}{\det(< \varphi_\mu, \varphi_\nu >)}.$$

Let

$$U_\lambda := \{Y \in y : \lambda \text{ is not in the spectrum of } \Delta_y\}$$

U_λ is open in Y, and the number $k(\lambda)$ of eigenvalues $< \lambda$ is constant in each component of U.
On $U_\mu \cap U_\lambda$(wlog $\mu > \lambda$), this yields a section s_μ of U_μ by

$$s_\mu(\varphi_1 \wedge ... \wedge \varphi_{k\mu}) = \sigma_\lambda \psi_1 \wedge ... \wedge \psi_{\ell(\lambda)} \otimes (D_{\varphi_{k(\lambda)+1}} \wedge ... \wedge D_{\varphi_{k(\mu)}}).$$

Thus

$$\sigma_\mu = \sigma_\lambda \otimes \det D_{(\lambda,\mu)},$$

where $E^1_{(\lambda,\mu)}$ corresponds to eigenvalues λ_n between λ and μ and

$$\det D_{(\lambda,\mu)} : \text{Det } E^1_{(\lambda,\mu)} \to \text{Det } E^2_{(\lambda,\mu)}$$

is the induced map. Thus

$$\begin{aligned} \|s_\mu\| &= \|s_\lambda\|^2_{\text{ind}} \, \| \det D_{(\lambda,\mu)}\|^2 \\ &= \|s_\lambda\|^2_{\text{ind}} \prod_{\lambda < \lambda_n < \mu} \lambda_n. \end{aligned}$$

Thus, the Quillen norm of the induced section s of $\det D$,

$$\|s\|^2_Q = \|s_\lambda\|^2_{\text{ind}} (\det D^* D_{|\lambda_n > \lambda}) \tag{2.9.10}$$

is independent of λ, if

$$\det D^* D_{|\lambda_n > \lambda} = \prod_{\lambda_n > \lambda} \lambda_n := \exp(-\zeta'_\lambda(0))$$

is again defined by zeta-function regularisation from

$$\zeta_\lambda(s) := \sum_{\lambda_n > \lambda} \frac{1}{\lambda_n^s}.$$

Namely, in that case, from the properties of zeta-functions

$$\det D^* D_{|\lambda_n > \lambda} = \det D^* D_{|\lambda_n > \mu} \prod_{\lambda < \lambda_n < \mu} \lambda_n. \tag{2.9.11}$$

□

In particular, we repeat the definition of the determinant of the differential operator D^*D

$$\det D^*D = \prod_{\text{eigenvalues}} \lambda_n := \exp(-\zeta'(0)).$$

In other words, the determinant as the product of the eigenvalues is defined by ζ-function regularization.
For the operator D, however, its determinant

$$\det D \in (\mathrm{Det}\ \ker D)^* \otimes \mathrm{Det}\ \ker D^*$$

is only defined as an element of the determinant line.
By (2.9.9), we can express the determinant of D^*D through the Quillen metric, and the transformation properties of such Quillen metrics will lead to anomaly cancellations in our partition function for strings.

The results of § 2.5 yield the determinant lines of the operators

$$P' = \bar{\partial}_{-1} \quad \text{and} \quad \bar{\partial}_0$$

considered there. Recall our notation that Σ is a Riemann surface of genus p with k boundary curves, and Σ' is its Schottky double which has genus $p' = 2p+k-1$. We then have, with our above boundary conditions

$$\begin{aligned} \text{(2.9.12)} \qquad \mathrm{Det}\ P'_{\Sigma'} &= \mathrm{Det}\ \bar{\partial}_{-1,\Sigma'} = (\ \mathrm{Det}\ker \bar{\partial}_{-1,\Sigma'})^* \otimes \mathrm{Det}\ \ker \bar{\partial}^*_{-1,\Sigma'} \\ &= \wedge^{\max}(H^0(\Sigma', K^2))^* \quad \text{for} \quad p' \geq 2, \\ &\text{by (2.5.21) - (2.5.25)} \end{aligned}$$

and

$$\text{(2.9.13)} \qquad \mathrm{Det}\ P'_{\Sigma} = \wedge^{\max}(H^0(\Sigma, K^2))^*, \ \text{cf Lemma 2.7.4.}$$

Thus

$$\text{(2.9.14)} \qquad \mathrm{Det}\ P'_{\Sigma} = \mathrm{Det}\ P'_{(\Sigma')} \otimes_{\mathbb{R}} \mathbb{C}.$$

Also

$$\text{(2.9.15)} \qquad \mathrm{Det}\ \bar{\partial}_{0,\Sigma'} = \mathbb{C} \otimes \wedge^{\max}(H^0(\Sigma', K))^*$$

and

(2.9.16) $\mathrm{Det}\ \bar{\partial}_{0,\Sigma} = i\mathbb{R} \otimes \wedge^{\max}(H^0(\Sigma, K))^*$, as established at the end of § 2.5.

Thus

$$\text{(2.9.17)} \qquad \mathrm{Det}\ \bar{\partial}_{0,\Sigma'} = \mathrm{Det}\ \bar{\partial}_{0,\Sigma} \otimes_{\mathbb{R}} \mathbb{C}.$$

Similarly, for the eigenvalues, we have the analogue of Lemma 2.7.4

Lemma 2.9.2. *With our above boundary conditions on Σ, $\bar\partial_n^*\bar\partial_n$ has the same eigenvalues on Σ and Σ', and the corresponding eigenspaces on the latter are the complexifications of the ones on the former.*

Proof. $\bar\partial_n^*\bar\partial_n$ is a real operator in the sense that it preserves the above decomposition of $n-$differentials ϕ into the "real" part $\phi_1(z) = \phi(z) + \bar\phi(i(z))$ and the "imaginary" part $\phi_2(z) = \phi(z) - \bar\phi(i(z))$.
Thus, if ϕ is an eigenform,

$$\bar\partial_n^*\bar\partial_n\phi = \lambda\phi,$$

then so are ϕ_1 and ϕ_2, with the same eigenvalue λ. Conversely, if $\psi = \psi_1 + i\psi_2$ is an n-differential on Σ with the above boundary conditions

$$\psi_2 = 0 = \psi_{1,y} \quad \text{on } \partial\Sigma \text{ (locally given as } \{y = 0\}),$$

that is an eigenform,

$$\bar\partial*_n\bar\partial_n\psi = \lambda\psi,$$

then it can be reflected to an eigenform on Σ'. □

For abbreviation, we now put

$$L_n := (\text{Det } \bar\partial_{1-n})^*. \tag{2.9.18}$$

Thus, on Σ', for $p' \geq 2$

$$L_1 = \wedge^{\max} H^0(\Sigma', K) \otimes \mathbb{C} \text{ by (2.9.15)} \tag{2.9.19}$$

$$L_2 = \wedge^{\max} H^0(\Sigma', K^2) \text{ by (2.9.12)}, \tag{2.9.20}$$

and in fact for all $n \geq 2$

$$L_n = \wedge^{\max} H^0(\Sigma', K^n).$$

Given a conformal metric $\rho^2 dz d\bar z$ on Σ', we obtain L^2-metrics on the L_n, and then also induced Quillen metrics $\|\cdot\|_{Q,n,\rho^2}$. The question how these metrics depend on $\rho^2 dz d\bar z$ is answered by

Lemma 2.9.3. *(Bost-Jolicoeur):*
Let $\rho_\lambda^2 dz d\bar z := e^{2\lambda}\rho^2 dz d\bar z$ for some functions $\lambda : \Sigma' \to \mathbb{R}$, be another conformal metric, confromally equivalent to $\rho^2 dz d\bar z$, and let

$$\vartheta := \frac{1}{2\pi i}\partial\bar\partial \log \rho^2$$

be the curvature form of the metric $\rho^2 dz d\bar z$. Then

$$\frac{\|\cdot\|^2_{Q,n,\rho_\lambda^2}}{\|\cdot\|^2_{Q,n,\rho^2}} = \exp\left(\frac{6n(n-1)+1}{6ni}\int_{\Sigma'}(\lambda\vartheta + \partial\lambda\wedge\bar\partial\lambda)\right). \tag{2.9.21}$$

This generalizes a computation of Quillen for the curvature form of the Quillen metric. For the proof, we refer to the literature, e.g. [9].

The following consequence of this lemma will be crucial for determining 26 as the unique dimension in which the two conformal anomalies encountered in our attempts to get rid of the conformal invariance by the Faddeev-Popov method will cancel each other:

Corollary 2.9.1. *The Quillen metric on $L_2^* \otimes L_1^{13}$ does not depend on the choice of conformal metric on Σ'.*

We can also easily write down the Quillen metrics on these line bundles. For that purpose, we choose bases $\omega_1, ..., \omega_{p'}$ (over $\mathbb{R}$ in case of the surface Σ of genus p with k boundary curves, and over $\mathbb{C}$ for its Schottky double Σ' of genus $p' = 2p+k-1$) of the space of holomorphic 1-forms (imaginary on the boundary for Σ)
and

$$q_1, ..., q_{3p'-3}$$

of the space of holomorphic quadratic differentials (again real on the boundary for Σ). These bases will depend on the conformal structure of Σ or Σ' as a Riemann surface, but not on the particular choice of a conformal metric. In other words, they are independent of a conformal factor in the metric.
A section of

$$L_1 = \wedge^{\max} H^0(\Sigma, K) \otimes i\mathbb{R}$$

is then given by

$$\sigma_1 = (\omega_1 \wedge ... \wedge \omega_{p'}) \otimes i, \tag{2.9.22}$$

with norm

$$\|\sigma_1\|^2_{\text{ind}} = \det(\omega_i, \omega_j) \int_\Sigma \rho^2 \frac{i}{2} dz \wedge d\bar{z} \ , \tag{2.9.23}$$

where $\rho^2 dz \otimes d\bar{z}$ is the conformal metric, and (ω_i, ω_j) is the L^2-product $\int_\Sigma \omega_i \wedge \bar{\omega}_j$ of the holomorphic 1-forms ω_i, ω_j. The Quillen norm of the section s_1 of L_1^* dual to the section σ_1 of L_1 is then given by

$$\|s_1\|^2_Q := \|s_1\|^2_{Q,1,\rho^2} = \det \bar{\partial}_0^* \bar{\partial}_0 \|s_1\|^2_{\text{ind}} = \frac{\det \bar{\partial}_0^* \bar{\partial}_0}{\|\sigma_1\|^2_{\text{ind}}} = \frac{\det \bar{\partial}_0^* \bar{\partial}_0}{\det(\omega_i, \omega_j) \int_\Sigma \rho^2} . \tag{2.9.24}$$

Similarly, a section of

$$L_2 = \wedge^{\max} H^0(\Sigma, K^2)$$

is given by

$$\sigma_2 = q_1 \wedge ... \wedge q_{3p'-3} \tag{2.9.25}$$

with norm

$$\|\sigma_2\|^2_{\text{ind}} = \det(q_i, q_j). \tag{2.9.26}$$

The Quillen norm of the section s_2 of L_2^* dual to σ_2 is given by

$$\|s_2\|^2_Q := \|s_2\|^2_{Q,2,\rho^2} = \frac{\det \bar{\partial}^*_{-1}}{\det(q_i, q_j)}. \tag{2.9.27}$$

Again, (q_i, q_j) is the L^2-product of the holomorphic quadratic differentials q_i, q_j (Weil-Petersson product, see above). In contrast to the L^2-product for holomorphic 1-forms, here we need to utilize the conformal metric $\rho^2 dz d\bar{z}$ on Σ.
For every (marked) Riemann surface Σ' without boundary (or Σ with boundary) we have the lines L_n, in particular, for $p' \geq 2$,

$$L_1 = \wedge^{\max} H^0(\Sigma', K) \otimes \mathbb{C}$$
$$L_2 = \wedge^{\max} H^0(\Sigma', K^2).$$

If we let Σ' vary in the Teichmüller space $T_{p'}$, we thus obtain line bundles over $T_{p'}$, with fibers L_n. By an abuse of notation, we denote these line bundles again by L_n. These line bundles turn out to be holomorphic line bundles, because if Σ' varies holomorphically in $T_{p'}$, so does the space of holomorphic n-differentials on Σ'.
The crucial result relating the various line bundles L_n is due to Mumford, and in the context of Polyakov's approach to string theory it was discovered, together with the implications worked out below, by Belavin-Knizhnik.

Theorem 2.9.1. *We have the isomorphism between holomorphic line bundles*

$$L_n \cong L_1^{6n(n-1)+1}.$$

It was shown by Bismut-Freed ([9]) that this is an isomorphism even as an isometry for the Quillen metrics.

In fact, these results not only hold on Teichmüller space, but also on moduli space although one needs to exercise some care in the interpretation as the moduli space does have singularities. There are two easy ways around this difficulty, however, namely

1. to consider only the regular locus of moduli space $M_{p'}$ or $M_{p,k}$; namely, since these spaces are noncompact anyway, actually quasi projective in the sense that they can be compactified to $M_{p'}$ or $\bar{M}_{p,k}$ by adding a divisor, it will not matter if we take a quasi projective subvariety of this compatification that is somewhat smaller than the moduli space itself, or alternatively,

2. to pass to a singularity free finite covering of the moduli space - which is known to exist - , and divide the corresponding integrals by the degree of the covering.

2.10 The partition function for the Bosonic string

The heuristic idea consists in trying to make sense of the formal path integral

$$\int dX dg\, e^{-D(X,g)} \tag{2.10.1}$$

where the integration is over the space of all mappings

$X : S \to \mathbb{R}^d$ with the prescribed Plateau type boundary conditions

and all metrics g on the differentiable surface S with boundary, and $D(X, g)$ is the Dirichlet integral as considered in § 2.1.
We recall the decomposition (2.4.2)

$$X = X_0 + \xi \tag{2.10.2}$$

with

$$\Delta_g X_0 = 0 \tag{2.10.3}$$

and

$$\xi_{|\partial S} = 0. \tag{2.10.4}$$

In other words, we decompose X into a harmonic function X_0 with the same boundary values ($X_{0|\partial S} = X_{|\partial S}$), and a function that vanishes on ∂S.
Since $D(X_0 + \xi, g) = D(X_0, g) + D(\xi, g)$,
because X_0 is harmonic (see § 2.2), we may write

$$\int dX\, e^{-D(X,g)} = \int dX_0\, e^{-D(X_0,g)} \int d\xi\, e^{-D(\xi,g)}. \tag{2.10.5}$$

In order to define $\int d\xi\, e^{-D(\xi,g)}$, as in § 1.3 our heuristic guide is an expansion of ξ in terms of eigenfunctions of Δ_g, and so we put

$$\int d\xi\, e^{-D(\xi,g)} = (\det \Delta_g)^{-\frac{d}{2}} \left(\int_S \sqrt{\det g} \right)^{\frac{d}{2}}, \tag{2.10.6}$$

where the d in the exponent arises as $\xi \in \mathbb{R}^d$ has d components.
The first observation is that (2.10.6) is completely independent of the boundary condition that X map ∂S onto $\Gamma = \Gamma_1 \cup \Gamma_2$ (of § 2.1); of course, this is not surprising as ξ itself satisfies the homogeneous boundary condition (2.10.4).
Thus, if we put

$$j_g(\Gamma) := j_g(\Gamma_1, \Gamma_2) := \int dX_0\, e^{-D(X_0,g)}, \tag{2.10.7}$$

with X_0 satisfying (2.10.3), we have

$$\int dX\, e^{-D(X,g)} = j_g(\Gamma)(\det \Delta_g)^{\frac{d}{2}} \left(\int_S \sqrt{\det g} \right)^{\frac{d}{2}} \tag{2.10.8}$$

from (2.10.5) - (2.10.7).
We now discuss the expression $j_g(\Gamma)$. It involves an integration over harmonic maps X_0 that map the boundary of S onto Γ in the required fashion. The key observation then is that such a harmonic map is already uniquely determined by its boundary values $X_{0|\partial S}$. Therefore, the integration reduces to an integration over the set of all permissible boundary values, by expressing the Dirichlet integral for a harmonic function in terms of its boundary values. This Dirichlet integral will require the structure of a Riemann surface Σ on S, but as discussed in § 2.1, it will not depend on the choice of conformal metric on Σ. If Σ is the unit disk D, then there even is an explicit formula for the Dirichlet integral of a harmonic function

$$Y : D \to \mathbb{R}$$

in terms of its boundary values

$$y := Y_{|\partial D},$$

due to J. Douglas, namely

$$D(Y) = \frac{1}{16\pi} \int_0^{2\pi}\int_0^{2\pi} \frac{(y(\varphi_1) - y(\varphi_2))^2}{\sin^2\left(\frac{\varphi_1 - \varphi_2}{2}\right)} d\varphi_1 d\varphi_2.$$

If we can set up the integration over the boundary values in such a manner that it is also independent of the choice of a conformal metric on Σ, then this part of our functional integral will not contribute a conformal anomaly. Integration over the boundary values now involves a one-dimensional domain as in § 1.3, and the key fact exploited there was that in that situation, in contrast to the two-dimensional case discussed in the remainder of the present chapter, there only remains a global constant factor after dividing out the action of the diffeomorphism group, and not an arbitrary positive function. Of course, we then have to fix that factor, and that is achieved as follows:
Let Σ be a Riemann surface with k boundary curves $\gamma_1, ..., \gamma_k$ with a conformal metric $g = \rho^2 dz d\bar{z}$. For $i = 1, ..., k$, we let

$$\eta^i : S^1 \to \gamma_i$$

be an oriented diffeomorphism. We then multiply the pulled back metric $(\eta^i)^* g$ on S^1 by a constant factor determined in such a manner that the length of S^1 w.r.t this metric equals the length of the component of our configuration Γ in $\mathbb{R}^d$ to which the boundary curve γ_i corresponds. In other words, we use the length scale in our target space in order to fix the intrinsic length scale of the string. For the rest of the discussion, it suffices to treat the case of one boundary component, and so we shall drop the index i in the sequel. We denote our rescaled metric on S^1 by

$$g_\eta = \lambda^2(t)dt^2.$$

As discussed in § 1.3, this induces an L^2-product on tangent vector fields V, W of S^1

$$(V, W) := \int_{S^1} \lambda^3(t)V(t)W(t)dt.$$

We want this product to correspond to a measure $d\varphi$ on $\mathcal{D}_+$, the group of oriented diffeomorphisms of S^1. Actually, there does not exist any nontrivial measure on $\mathcal{D}_+$, but if we pass to the larger group of all oriented monotone self maps of S^1, then such a measure exists as one knows from the discussion of the Wiener integral. In fact, the appropriate regularity class is the Sobolev space $H^{\frac{1}{2}}(S^1)$ of those self maps x of S^1 for which $\dot{x}x$ exists. Namely, these are precisely the maps whose harmonic extension X has finite Dirichlet integral as $\int_\Sigma DX \cdot DX = -\int_\Sigma \Delta X \cdot X + \int_{\partial\Sigma} \dot{x}x$. W.r.t. this measure, we now put

$$j_\Sigma(\Gamma) := \int e^{-E(x_0\circ\varphi, g_0)} d\varphi, \tag{2.10.9}$$

with the following notations: g_0 is some conformal metric on Σ, $x_0 : \gamma \to \Gamma$ is some oriented diffeomorphism, and

$$E(x_0 \circ \varphi, g_0)$$

is the Dirichlet integral of the harmonic extension

$$H(x_0 \circ \varphi) : \Sigma \to \mathbb{R}^d$$

of $x_0 \circ \varphi : \partial\Sigma \to \mathbb{R}$, i.e.

$$E(x_0 \circ \varphi, g_0) = D(H(x_0 \circ \varphi), \Sigma),$$

in the notation of § 2.1, recalling that the Dirichlet integral is conformally invariant.

In order to make $j_\Sigma(\Gamma)$ well defined, we need to check that it is independent of the choices of η, x_0 and g_0. It is independent of the choices of η and x_0, because any two choices differ by an oriented diffeomorphism of S^1, and we are integrating over all such diffeomorphisms.

If we change g_0 by a diffeomorphism ψ, then because of

$$D(X, \psi^* g_0) = D(X \circ \psi^{-1}, g_0),$$

we see that this can be accounted for by changing the boundary values by the inverse diffeomorphisms, and so again this is accounted for by our functional

integral. Finally, if we multiply g_0 by a conformal factor, then again we change the induced metric on S^1, but up to the choice of a global factor which we have already fixed, this again can be accounted for by the action of the diffeomorphism group of S^1 as discussed above and in § 1.3. An older reference for the discussion of the boundary values is O. Alvarez [2].
Thus, we see that $j_\Sigma(\Gamma)$ as defined in (2.10.9) is indeed independent of all choices involved; it does depend, however, on the conformal structure Σ. Thus, in the sequel, we are permitted to write $j_\Sigma(\Gamma)$ in place of $j_g(\Gamma)$ in (2.10.7); in fact, denoting the coordinates on Teichmüller space by t, we shall prefer to write

$$j_t(\Gamma).$$

We thus need to define the integral

$$\int dg\; j_t(\Gamma)\left(\frac{\det \Delta_g}{\int\sqrt{\det g}}\right)^{\frac{-d}{2}}.$$

In § § 2.3, 2.4, we have decomposed the tangent space to the space of all metrics on S into those infinitesimal directions that are Lie derivatives of the given metric g in the direction of a vector field, i.e. infinitesimal diffeomorphisms, into the conformal changes of the metric, and into the Teichmüller directions. As everything is invariant under the action of the diffeomorphism group as verified in § 2.2, we would like to divide out that action and discard the integration over the infinitesimal diffeomorphisms. According to the Faddeev-Popov procedure , this leaves us with a determinant, and we identified that determinant in (2.6.17). From (2.7.19) (see the discussion before (2.7.22)), we may identify that determinant with the square of the one of $\bar\partial_{-1}$; in other words, we get the factor

$$(\det \bar\partial^*_{-1}\bar\partial_{-1}),$$

and that expression has been defined in § 2.6.
Altogether, we are left with the following formal expression for the partition function

$$Z(\Gamma) = \int\limits_{Conf(s)} d\sigma \int\limits_{T_{p,k}} d(WP) j_t(\Gamma)\left(\frac{\det \Delta_g}{\int\sqrt{\det g}}\right)^{\frac{-d}{2}} (\det \bar\partial^*_{-1}\bar\partial_{-1}) \tag{2.10.10}$$

where $d\sigma$ is the - formal - measure on the space of conformal factors, while

$$\int\limits_{T_{p,k}} d(WP)$$

represents an integral over Teichmüller space w.r.t the Weil-Petersson measure. Employing the above basis

$$q_1, \ldots, q_{3p'-3}$$

of holomorphic quadratic differentials, i.e. tangent directions to $T_{p,k}$, we may write

$$d(WP) = \frac{dt_1 \wedge .. \wedge dt_{3p'-3}}{\det(q_i, q_j)} \tag{2.10.11}$$

where t_j is the coefficient in the direction q_j. The above expression for $Z(\Gamma)$ still encounters two problems:

1. the measure $d\sigma$ on the infinite dimensional space of conformal factors is not defined
2. the volume of $T_{p,k}$ w.r.t the Weil-Petersson measure is infinite, as this measure is invariant under the infinite group $\Gamma_{p,k}$, the modular group acting discretely by isometries on $T_{p,k}$.

This, however, is now easily handled:
With (2.10.11) inserted in (2.10.10), we simply recall (2.9.24) and (2.9.27) to rewrite $Z(\Gamma)$ as

$$Z(\Gamma) = \int\limits_{Conf(S)} d\sigma \int\limits_{T_{p,k}} dt_1 \wedge ... \wedge dt_{3p'-3}\; j_t(\Gamma) \|s_1\|_Q^{-d} \|s_2\|_Q^2 \det(\omega_i, \omega_j)^{\frac{-d}{2}}.$$

Now the important point is that for $d = 26$, we may apply Corollary 2.9.1 to obtain that for that particular dimension, the integrand does not depend on the conformal factor σ anymore. Therefore, for $d = 26$, we may simply discard the integration over $Conf(S)$ and redefine the partition function as

$$Z(\Gamma) := \int\limits_{T_{p,k}} dt_1 \wedge ... \wedge dt_{3_{p'}-3} \|s_1\|_Q^{-26} \|s_2\|_Q^2 \det(\omega_i, \omega_j)^{-13}.$$

This solves the first problem. The second one is handled similarly. Namely, by Theorem 2.9.1 and the discussion following it, $s_1^{-13} \otimes s_2$ is a section of the line bundle

$$L_1^{*-13} \otimes L_2^*$$

that is trivial on the moduli space $M_{p,k}$. Therefore, the integrand is invariant under the action of the modular group $\Gamma_{p,k}$, and so, we may redefine $Z(\Gamma)$ once more as an integral over $M_{p,k}$ in place of $T_{p,k}$:

$$Z(\Gamma) := \int\limits_{M_{p,k}} dt_1 \wedge ...dt_{3_{p'}-3} \|s_1\|_Q^{-26} \|s_2\|_Q^2 \det(\omega_i, \omega_j)^{-13}.$$

In fact, $s_1^{-13} \otimes s_2$ is a holomorphic section of the trivial line bundle with a flat connection, $L_1^{-13} \otimes L_2^*$, and so it may simply be considered as a holomorphic function f on $M_{p,k}$. Thus,

$$Z(\Gamma) = \int\limits_{M_{p,k}} dt_1 \wedge ... \wedge dt_{3_{p'}-3} \|f(t)\|^2 \det(\omega_i, \omega_j)^{-13}.$$

Thus, in dimension $d = 26$, we have now arrived at a mathematically well defined expression that can be considered as the functional integral

$$\int\int dgdX\, e^{-D(X,g)}$$

after dividing out all invariances. This is the **fundamental result for the bosonic string**.

2.11 Some physical aspects

For the physical theory, we should not employ a Riemann surface with a positive definite metric, but rather a surface with a Lorentzian metric. As we have already thoroughly discussed the role of the domain metric, in the present §, we only consider the simplest case, namely a strip

$$\{(\sigma,\tau) : 0 \leq \sigma \leq \ell, \tau \in \mathbb{R}\},$$

with the metric tensor

$$\begin{pmatrix} 1 & 0 \\ 0 & -1 \end{pmatrix}.$$

Likewise, the target space now is Minkowski space $\mathbb{R}^{1,25}$, with Lorentz metric $\eta_{\mu\nu}$ of signature $(-,+,...,+)$. For the fields, we write

$$\dot{X}^\mu(\sigma,\tau) := \frac{\partial}{\partial\tau}X^\mu(\sigma,\tau)$$
$$X'^\mu(\sigma,\tau) := \frac{\partial}{\partial\sigma}X^\mu(\sigma,\tau).$$

The action is

(2.11.1) $$S = \frac{1}{4\pi\alpha'}\int_0^\ell d\sigma \int d\tau((\dot{X}^\mu)^2 - (X'^\mu)^2).$$

If we impose either the Neumann boundary conditions

(2.11.2) $$X'^\mu(0,\tau) = 0 = X'^\mu(\ell,\tau) \quad \text{(open string)}$$

or the periodic boundary conditions

(2.11.3) $$X^\mu(0,\tau) = X^\mu(\ell,\tau),\ X'^\mu(0,\tau) = X'^\mu(\ell,\tau) \quad \text{(closed string)},$$

the equations of motion are the wave equations

(2.11.4) $$\Box X^\mu(\sigma,\tau) = \left(\frac{\partial^2}{\partial\sigma^2} - \frac{\partial^2}{\partial\tau^2}\right)X^\mu(\sigma,\tau) = 0 \quad \text{for all } \mu.$$

The general solution is

(2.11.5) $$X^\mu(\sigma,\tau) = X_L^\mu(\sigma+\tau) + X_R^\mu(\sigma-\tau).$$

The conjugate momenta are

(2.11.6) $$P^\mu(\sigma,\tau) = \frac{1}{2\pi\alpha'}\dot{X}^\mu(\sigma,\tau) = \frac{1}{2\pi\alpha'}\left(X'^\mu_L - X'^\mu_R\right),$$

with

$$X'_{L,R}(y) = \frac{d}{dy}X_{L,R}(y).$$

With

$$x^\mu(\tau) := \frac{1}{\ell}\int_0^\ell d\sigma X^\mu(\sigma,\tau) \quad \text{center of mass}$$

$$p^\mu(\tau) := \int_0^\ell d\sigma P^\mu(\sigma,\tau) = \frac{1}{2\pi\alpha'}\int_0^\ell d\sigma \dot{X}^\mu(\sigma,\tau),$$

the Fourier expansion of the general solution of (2.11.4), (2.11.2) is

(2.11.7) $$X^\mu(\sigma,\tau) = x^\mu + \frac{2\pi\alpha'}{\ell}p^\mu\tau + i(2\alpha')^{\frac{1}{2}} \sum_{n\in\mathbb{Z}\setminus\{0\}} \frac{1}{n}\alpha_n^\mu \exp\left(-\frac{\pi\mu n\tau}{\ell}\right)\cos\left(\frac{\pi\mu\sigma}{\ell}\right).$$

For X^μ to be real, we must have

(2.11.8) $$\alpha^\mu_{-n} = \overline{(\alpha^\mu_n)}.$$

Likewise, the solution of (2.11.4), (2.11.3) is given by

(2.11.9) $$X^\mu(\sigma,\tau) = x^\mu + \frac{2\pi\alpha'}{\ell}p^\mu\tau + i\left(\frac{\alpha'}{2}\right)^{\frac{1}{2}} \sum_{n\in\mathbb{Z}\setminus\{0\}} \frac{1}{n}\alpha_n^\mu \exp\left(-\frac{2\pi\mu n(\sigma+\tau)}{\ell}\right) + i\left(\frac{\alpha'}{2}\right)^{\frac{1}{2}} \sum_{n\in\mathbb{Z}\setminus\{0\}} \frac{1}{n}\tilde{\alpha}_n^\mu \exp\left(\frac{2\pi\mu n(\sigma-\tau)}{\ell}\right),$$

with the reality conditions

(2.11.10) $$\alpha^\mu_{-n} = \overline{(\alpha^\mu_n)},\ \tilde{\alpha}^\mu_{-n} = \overline{(\tilde{\alpha}^\mu_n)}.$$

Quantization leads to the commutation relations

(2.11.11) $$[X^\mu(z), P^\nu(z')] = i\eta^{\mu\nu}\delta(z,z').$$

Of course, X^μ and P^μ now are operators. In terms of the Fourier coefficients, in the expansion (2.11.7), we obtain

(2.11.12) $$[x^\mu, p^\nu] = i\eta^{\mu\nu}$$

(2.11.13) $$[\alpha^\mu_m, \alpha^\nu_n] = m\eta^{\mu\nu}\delta_{m,-n}.$$

Writing for $m > 0$

(2.11.14) $$\alpha^\mu_m = \sqrt{m}a, \alpha^\mu_{-m} = \sqrt{m}\, a^+, \quad (a = a^\mu_m)$$

(2.11.13) becomes the harmonic oscillator algebra with

$$[a, a^+] = 1.$$

Thus, we obtain a harmonic oscillator for each coordinate direction μ and each Fourier index m. Starting from the state $|0,0>$, the ground state of a single string with zero momentum (not to be confused with the zero-string vacuum state), we obtain the states $|0,k>$ with

(2.11.15)
$$p^\mu|0,k> = k^\mu|0,k> \quad \text{(eigenstates of the center-of-mass momenta)}$$
(2.11.16)
$$\alpha^\mu_m|0,k> = 0 \quad \text{for } m > 0 \quad \text{(annihilation by the lowering operators)}$$

as well as the general states obtained by applying the raising operators $\alpha^\mu_{-m} = \sqrt{m}\ a^{\mu+}_m$ to the states $|0,k>$. These states are denoted by $|N,k>$, with $N = (N^\mu_n)$ being the set of occupation numbers for each mode (μ, n). While the center-of-mass momenta are simply the degrees of freedom of a point particle as described by (x^μ, p^μ), the general states represent an infinite number of internal degrees of freedom. Every choice of the occupation numbers N^μ_n represents a different particle or spin state in space-time. We now make the standard analytic continuation from a Minkowski metric to a Euclidean one on the world sheet. For that purpose, we put

(2.11.17)
$$\sigma^1 = \sigma,\ \sigma^2 = i\tau,$$
$$z = \sigma^1 + i\sigma^2,\ \bar{z} = \sigma^1 - i\sigma^2.$$

We write the fields as $X^\mu(z,\bar{z})$, and the action now is

(2.11.18) $$S = \frac{1}{2\pi\alpha'}\int d^2z\, \partial X^\mu\, \bar{\partial} X^\mu,$$

and the equation of motion is the Laplace equation

(2.11.19) $$\partial\bar{\partial}X^\mu = 0\ (= \bar{\partial}\partial X^\mu).$$

This means that ∂X^μ is holomorphic, and so we write it as $\partial X^\mu(z)$, while $\bar\partial X^\mu$ is antiholomorphic and written as $\bar\partial X^\mu(\bar z)$. The expansions then take the form

$$\partial X^\mu(z) = -i\left(\frac{\alpha'}{2}\right)^{\frac{1}{2}} \sum_{m\in\mathbb{Z}} \frac{\alpha_m^\mu}{z^{m+1}} \quad \text{(called left-moving part)}$$

$$\bar\partial X^\mu(\bar z) = -i\left(\frac{\alpha'}{2}\right)^{\frac{1}{2}} \sum_{m\in\mathbb{Z}} \frac{\tilde\alpha_m^\mu}{\bar z^{m+1}} \quad \text{(called right-moving part).} \tag{2.11.20}$$

After integration

$$X^\mu(z,\bar z) = x^\mu - i\frac{\alpha'}{2}p^\mu \log\|z\|^2 + i\left(\frac{\alpha'}{2}\right)^{\frac{1}{2}} \sum_{m\in\mathbb{Z}\setminus\{0\}} \frac{1}{m}\left(\frac{\alpha_m^\mu}{z^m} + \frac{\tilde\alpha_m^\mu}{\bar z^m}\right). \tag{2.11.21}$$

We now want to study so-called compactifications of some of the directions of our 26-dimensional spacetime.
For that purpose, we consider $\mathbb{R}^{1,d}$ (in our case $D = d+1 = 26$, but at the moment, d can be arbitrary) with some Lorentzian metric

$$g^d_{\mu\nu}dx^\mu dx^\nu.$$

We wish to make the last coordinate periodic:

$$x^d \cong x^d + 2\pi R,$$

i.e. the last coordinate direction now represents a circle of length $2\pi R$. R is called radius of the compactification.
We shall let the indices i,j range from 0 to $d-1$, whereas μ,ν range from 0 to $D-1$. We rewrite the above metric as

$$g^d_{\mu\nu}dx^\mu dx^\nu = g_{ij}dx^i dx^j + g_{dd}(dx^d + A_i dx^i)^2,$$

where we request that now g_{ij}, g_{dd} and A_i only depend on the coordinates $x^0, \ldots, x^{d-1}$.
Reparametrizations

$$x'^d = x^d + \vartheta(x^i)$$

now change A_i to

$$A_i' = A_i - \frac{\partial\vartheta}{\partial x^\mu}.$$

Thus, A transforms as a gauge field. This is the original idea of Kaluza.
We now assume for simplicity

$$g_{dd} = 1,$$

and we consider a massless scalar φ (i.e. a solution of the wave equation) in D dimensions. Because of the periodicity of the x^d direction, its Fourier expansions is

$$\varphi(x^\mu) = \sum_{n\in\mathbb{Z}} \varphi_n(x^j) \exp\left(in\, {x^d}/{R}\right)$$

and the wave equation ($\partial_\mu \partial^\mu \varphi = 0$) becomes

$$\partial_\mu \partial^\mu \varphi_n(x^j) = \frac{n^2}{R^2}\varphi_n(x^j).$$

The momentum in the x^d-direction is quantized.

$$p_d = \frac{n}{R},$$

and if $p_d \neq 0$, i.e. $n \neq 0$, the mass is

$$m^2 = p_d p_d = \frac{n^2}{R^2}.$$

If instead of a point particle, we have a closed string, then the periodicity yields an integer winding number $w \in \mathbb{Z}$:

$$X^d(\sigma + \ell) = X^d(\sigma) + 2\pi w R.$$

One now computes the mass as

$$m^2 = \frac{n^2}{R^2} + \frac{w^2 R^2}{\alpha'^2} + \frac{2}{\alpha'}(N + n\tilde{N} - 2),$$

with the constraint

$$nw + N - \tilde{N} = 0.$$

Here, N and $\tilde{N}$ are the occupation numbers for the oscillators, and the summand -2 comes from the zero-point energy, a regularization required to make the above expansions converge. If we translate things back to the domain of the string, going around the string yields the total change of X via

$$2\pi w R = \oint (dz\partial X^d + d\bar{z}\bar{\partial} X^d) = z\pi \left(\frac{\alpha'}{2}\right)^{\frac{1}{2}} (\alpha_0 - \tilde{\alpha}_0),$$

see (2.11.20). We also have

$$p^{25} = \frac{1}{2\pi\alpha'} \oint (dz\partial X^d - d\bar{z}\bar{\partial} X^d) = \frac{1}{(2\alpha')^{\frac{1}{2}}}(\alpha_0 + \tilde{\alpha}_0).$$

These two equations yield

$$p_L^{25} := \left(\frac{2}{\alpha'}\right)^{\frac{1}{2}} \alpha_0 = \frac{n}{R} + \frac{wR}{\alpha'}$$

$$p_R^{25} := \left(\frac{2}{\alpha'}\right)^{\frac{1}{2}} \tilde{\alpha}_0 = \frac{n}{R} - \frac{wR}{\alpha'}.$$

As $R \to \infty$, the masses of the winding states go to infinity, while the contributions from the moments corresponding to the compactified direction tend to a continuous spectrum. As $R \to 0$, the roles are interchanged, and this suggests to study the duality transformation

$$R \to R' = \frac{\alpha'}{R}\ , \ n \leftrightarrow w$$

that leaves the spectrum invariant. This is called T-duality. In particular, as $R \to 0$, we get a continuum of states from the winding states, and in this sense, no degree of freedom is lost in the $R \to 0$ limit. This seems to be different for open strings, as here, we do not have topological winding numbers. This changes, however, if we constrain the end points of the open string to lie on a fixed hyperplane, i.e. if we impose Dirichlet instead of Neumann boundary conditions in the x^d-direction.
In our above notation, the duality is expressed as

$$P_L^{25} \to P_L^{25}\ , \ P_R^{25} \to -P_R^{25},$$

or

$$X^{25}(z,\bar{z}) = X_L^{25}(z) + X_R^{25}(\bar{z}) \to X'^{25}(z,\bar{z}) = X_L^{25}(z) - X_R^{25}(\bar{z}).$$

If ∂_n denotes a normal derivative at the boundary, and ∂_t is a tangential one, then

$$\partial_n X^{25} = -i\partial_t X'^{25},$$

and so T-duality converts our original Neumann boundary condition for the open string in the compactified direction into a Dirichlet boundary condition for the dual string. Thus, after applying the duality transformation, the X^{25} coordinate of the endpoints of the open string is fixed. In other words, the boundary of the string has to lie on a fixed hyperplane.
We had seen above that compactification creates a gauge field A_i from the space time metric, and such a gauge field varies under coordinate transformations in the not compactified directions. Similarly, the hyperplane here is affected by such coordinate transformations, and it becomes a dynamical object, a Dirichlet membrane or D-brane .

Bibliography

[1] S. Albeverio, J. Jost, S. Paycha and S. Scarlatti, A mathematical introduction to string theory, Cambridge Univ. Press, 1997.

[2] O. Alvarez, Theory of strings with boundaries, Nucl. Phys. B. 216, 125 - 184, 1983.

[3] W. Baily, On the moduli of Jacobian varieties, Ann. Math. 71, 303-314, 1960.

[4] W. Baily, A. Borel, Compactification of arithmetic quotients of bounded symmetric domains, Ann. Math. 84, 442-528, 1966.

[5] L. Bers, Spaces of degenerating surfaces, in: Discontinuous groups and Riemann surfaces, Ann. Math. Studies 79, Princeton Univ. Press, 1974.

[6] J. Bismut, D. Freed. The analysis of elliptic families, I, Comm. Math. Phys. 106, 159 - 176, 1986.

[7] R. Courant, Dirichlet's principle, conformal mapping, and minimal surfaces, Interscience, New York, 1950.

[8] P. Deligne, D. Mumford, The irreducibility of the space of curves of given genus, Publ. Math IHES 36, 75-110, 1969.

[9] D. Freed, On determinant line bundles, in: S. T. Yau (ed.), Mathematical aspects of string theory, pp. 189 - 238, World Scientific, Singapore, 1987.

[10] M. Green, J. Schwarz, E. Witten, Superstring theory I, Cambridge Univ. Press, 1995.

[11] M. Green, J. Schwarz, E. Witten, Superstring theory II, Cambridge Univ. Press, 1995.

[12] L. Habermann, J. Jost, Riemannian metrics on Teichmüller space, man. math. 89, 281-306, 1996.

[13] L. Habermann, J. Jost, Metrics on Riemann surfaces and the geometry of moduli spaces, in: J.-P. Bourguignon, P. de Bartolomeis, M. Giaquinta (eds.), Geometric Theory of Singular Phenomena in Partial Differential Equations, Cortona 1995, Cambridge Univ. Press, 53-70, 1998.

[14] E. d'Hoker, D. H. Phong, Multiloop amplitudes for the bosonic Polyakov string, Nucl. Phys. B 269, 205 - 234, 1986.

[15] E. d'Hoker, D.H. Phong, The geometry of string perturbation theory, Rev. Mod. Phys. 60, 917 - 1065, 1988.

[16] J. Jost, Compact Riemann surfaces, Springer, 1997.

[17] J. Jost, Two-dimensional geometric variational problems, Wiley-Interscience, 1991.

[18] J. Jost, Partielle Differentialgleichungen, Springer, 1998.

[19] J. Jost, Riemannian geometry and geometric analysis, 2nd edition, Springer, 1998.

[20] J. Jost, Minimal surfaces and Teichmüller theory, in: S.T. Yau (ed.), Tsing Hua Lectures on Geometry and Analysis, International Press, Cambridge, Mass., 149-211, 1997.

[21] J. Jost, M. Struwe, Morse-Conley theory for minimal surfaces of varying topological type, Invent. Math. 102, 465 - 499, 1990.

[22] D. Lüst, S. Theisen, Lectures on string theory, Lecture notes in physics, Springer, 1989.

[23] H. Masur, The extension of the Weil-Petersson metric to the boundary of Teichmüller space, Duke Math. J. 43, 623-635, 1976.

[24] C. Morrey, Multiple integrals in the calculus of variations, Springer, Berlin, 1966.

[25] D. Mumford, Stability of projective varieties, L'Enseignement Math. 23, 39-110, 1977.

[26] Y. Namikawa, Toroidal compactification of Siegel spaces, Springer Lect. Notes Math. 812, 1980.

[27] J. Polchinski, String theory, Vol. I, Cambridge Univ. Press, 1998.

[28] D. Quillen, Determinants of Cauchy-Riemann operators over a Riemann surface, Funct. Anal. Appl. 19, 31 - 34, 1985.

[29] I. Satake, On the compactificaiton of the Siegel space, J. Indian Math. Soc. 20, 259-281, 1956.

[30] A. Tromba, Teichmüller theory in Riemannian geometry, Birkhäuser, 1992.

[31] S. Wolpert, Noncompleteness of the Weil-Petersson metric for Teichmüller space, Pacific J. Math. 61, 513-576, 1975.

[32] W. Ziemer, Weakly differentiable functions, Springer, 1989.

Index